AF454338

Plant Breeding
Related Legislations at a Glance

The Author

 **Dr. Phundan Singh [b. 1946]** hails from a reputed agricultural family of Western Uttar Pradesh [Village-Sakauti, Tehsil- Mawana, District- Meerut]. He graduated in Agriculture from Agra University, Agra and obtained his Master and Doctorate degrees from Kanpur University, Kanpur. He has throughout a brilliant academic record. His area of specialization is Plant Breeding and Genetics. Dr. P. Singh has total experience of more than 40 years in the field of Genetics and Plant Breeding. He has served the Central Institute for Cotton Research, Nagpur for about 32 years in various capacities such as Scientist, Senior Scientist, Principal Scientist, Head of Division and Director. He has 300 publications to his credit. He has participated and presented 80 research papers in various National and International Seminars, Symposia and Conferences He has authored more than 90 books on Genetics, Plant Breeding, Seed Technology, Biotechnology, *etc.* He has also authored 18 technical bulletins on various aspects of cotton and contributed 14 chapters in various books. Dr. P. Singh is a member of several Scientific Societies, referee of different journals, Expert member of Selection Committees and M.Sc. and Ph.D. Examiner in different Agricultural Universities. The Institute was recognized as the referral Laboratory for detection of Bt. gene and received the best Annual Report Award when he was the Director. He has visited several countries such as Russia, Belarus, Ukraine, Crimea, Uzbekistan, Canada, USA, England, Austria and Germany.

Plant Breeding
Related Legislations at a Glance

Phundan Singh

Former Director
ICAR-Central Institute for Cotton Research,
Nagpur – 440 010
Maharashtra, India

2022

Daya Publishing House®

A Division of

Astral International Pvt. Ltd.
New Delhi – 110 002

ISBN : 9789354615252

Publisher's Note:

Every possible effort has been made to ensure that the information contained in this book is accurate at the time of going to press, and the publisher and author cannot accept responsibility for any errors or omissions, however caused. No responsibility for loss or damage occasioned to any person acting, or refraining from action, as a result of the material in this publication can be accepted by the editor, the publisher or the author. The Publisher is not associated with any product or vendor mentioned in the book. The contents of this work are intended to further general scientific research, understanding and discussion only. Readers should consult with a specialist where appropriate.

Every effort has been made to trace the owners of copyright material used in this book, if any. The author and the publisher will be grateful for any omission brought to their notice for acknowledgment in the future editions of the book.

Published by : **Daya Publishing House®**
A Division of
Astral International Pvt. Ltd.
– ISO 9001:2015 Certified Company –
4736/23, Ansari Road, Darya Ganj
New Delhi-110 002
Ph. 011-43549197, 23278134
E-mail: info@astralint.com
Website: www.astralint.com

— Dedicated to —

Dr. Saksham Singh [Daughter in-law]
Er. Rajeev Singh [Son]
Ms. Pavitra Singh [Daughter]

Preface

There are various legislations related to plant breeding and seed technology. Such legislations can be broadly classified into four groups, *viz..* seed legislations, varieties legislations, biodiversity legislations, biosafety legislations, *etc.* However, the information on all these legislations is not available from any single source and students have to go through voluminous literature to cover their course requirements. Present book has been designed to provide comprehensive information about all these legislation in question and answer form in one compact volume.

This book covers all plant breeding related legislations in its 16 chapters. The book has been divided in to five sections. Section 1 deals with seed related legislations consisting of 8 chapters. Section 2 covers deals with varieties related legislations in 3 chapters, Section 3 deals with biodiversity legislations in 3 chapters, Section 4 deals with biosafety legislations in two chapters and section 5 deals with miscellaneous topics consisting of 3 chapters. Each chapter has been presented point wise and step by step for easy grasping of students. Glossary of technical terms is presented at the end of each chapter. Some useful references are given at the end for gathering further details.

The information contained in this book has been gathered from various published sources and Internet websites. Attempts have been made to provide latest information even then some valuable information might have been missed.

The cooperation extended by my wife Mrs. Jaswanti Singh during preparation of this book is highly appreciable. Hope this Volume would be useful to the students, researchers, teachers and seed growers engaged in the field of plant breeding and seed technology. Constructive suggestions of students and teachers are invited for further improvement of this book.

Phundan Singh

Contents

Section IV: Biosafety Legislations

Section V: Miscellaneous Topics

Seed Related Legislations

Chapter 1

Introductory Chapter

Q1. What are various plant breeding related Legislations?

Ans. Various legislations related to plant breeding and genetics approved and implemented by the Government of India so far can be broadly divided into four major groups, *viz..*, seed legislations, varieties legislations, biodiversity legislations and biosafety legislations (Table 1-1).

TABLE 1-1: List of Plant Breeding Related Legislations

Sl.No.	Areas of Legislations	List of Legislations
1	Seed Legislations	Seed Act, 1966
		Seed Rules, 1968
		Seed Amendment Rules, 1972
		Seed Amendment Rules, 1973
		Seed Amendment Rules, 1974
		Seed Amendment Rules, 1981
		New Seed Policy, 1988
		Plants, Fruits and Seed Orders, 1989
		National Seed Policy, 2002
		Seed Act, 2004
2	Varieties Legislations	Union for Protection of Plant Varieties (UPOV) Act, 1978
		UPOV Act 1991
		Protection of Plant Varieties and Farmers Rights, 2001
		Protection of Plant Varieties Rules, 2003
		Farmers' Rights Act 2001
		Plant Breeders' Rights Act

Sl.No.	Areas of Legislations	List of Legislations
3	Biodiversity Legislations	Convention on Biodiversity Indian Biodiversity Legislation
4	Biosafety Legislations	Cartagena Protocol on Biosafety Biosafety Regulations for Transgenic Plants
5	Miscellaneous Legislations	Plant Quarantine order, 2003 Intellectual Property Rights.

Q2. Describe briefly various seed related Legislations.

Ans. In India, the Seed Act was first introduced in 1966 which is called as Seeds Act 1966. This Act was amended in 1972 which is known as the Seeds amended Act 1972. Subsequently, the seed act was thoroughly revised and the revised act was approved by the Government of India in September 2004. The revised Seeds Act is known as Seeds Act 2004 which came into force with effect from January 2005. It replaces the Seeds Act, 1966. The Indian seed Act has undergone several changes over the years. Seed legislations were formulated and implemented with the following objectives.

(i) To ensure availability of quality seeds to the farmers.

(ii) To regulate production and marketing of planting seed through various government organizations such as national Seeds Corporation, State Seeds Corporations, State Seed Certification Agencies, State Agricultural Universities and Central Agricultural Universities.

Q3. Give a brief account of various varieties related Legislations.

Ans. Various plant breeding legislations related to plant varieties have been formulated and implemented both at International and national levels. International legislations include Union for Protection of Plant Varieties such as UPOV Act, 1978 and UPOV Act, 1991. In India, Protection of Plant Varieties and Farmers' Rights Act was approved by the government of India in 2001. Other legislations related to plant varieties include Plant Breeders Rights and farmers Rights. In India, these have been covered by Protection of Plant Varieties and Farmers' Rights Act. Varieties related legislations were formulated and implemented with the following objectives.

(i) To provide legal rights to the original plant breeders to regulate production and marketing of his varieties. In other words, original plant breeder can take advantage of varieties developed by him/her.

(ii) To provide legal rights to the farmers to use protected plant varieties for commercial crop production. He can use seed of his crop for raising further crop or he can exchange seed of such varieties with other farmers.

(iii) To make use of protected plant varieties by plant breeders for developing new superior crop varieties.

Q4. Describe briefly biodiversity related legislations.

Ans. The Convention on Biological Diversity is also known as the Biodiversity Convention. It is an international treaty that was adopted in Rio de Janeiro in June 1992. The Convention was opened for signature at the Earth Summit in Rio de Janeiro on 5 June 1992 and entered into force on 29 December 1993. The biodiversity agreement has been signed by 189 countries. In India the Biological Diversity Act [also called biodiversity act] was passed by the Central Government in 2002.Biodiversity related legislations have the following three main goals:

   (i) Conservation of biological diversity (or biodiversity):

  (ii) Sustainable use of its components; and

 (iii) Fair and equitable sharing of benefits arising from genetic resources.

Q5. Explain in brief biosafety related legislations.

Ans. The Cartagena Protocol on Biosafety is the first international agreement to regulate the trans-boundary movements of genetically engineered (GE) organisms. The Biosafety Protocol is a subsidiary agreement to the UN Convention on Biological Diversity (CBD), which was signed by over 150 governments at the Rio Earth Summit in 1992. The Biosafety Protocol seeks to protect biological diversity from the potential risks posed by living modified organisms [LMOs] resulting from modern biotechnology.

Q6. Describe briefly important events related to Plant Breeding Legislations in India

Ans. Seed Act refers to the legal procedures approved by the government for production and marketing of seeds. The main objective of the Seeds Act is to ensure availability of quality seeds to farmers. In India, the Seed Act was first introduced in 1966 which is called as Seeds Act 1966. This Act was amended in 1972 which is known as the Seeds amended Act 1972. Subsequently, the seed act was thoroughly revised and the revised act was approved by the Government of India in September 2004. The revised Seeds Act is known as Seeds Act 2004 which came into force with effect from January 2005. It replaces the Seeds Act, 1966. The Indian seed Act has undergone several changes over the years. Important events related to plant breeding related legislations are presented in Table 1.2.

TABLE 1-2: Important Events Related to Plant Breeding Legislations in India

Year	Important Event	Remarks/Brief Description
1966	The first Indian Seed Act was formulated	To provide good quality planting seeds to the farmers.
1968	Seed Rules were framed to implement various legislations given Seed Act 1966.	Enactment of Seed Act 1966.
1972	The Seed Act 1966 was amended known as seed amendment Rules 2002.	The Jute seeds were included to the Seeds Act, establishment of Seed Certification Board and fixing of minimum standards.

Year	Important Event	Remarks/Brief Description
1973	The Seed Act 1966 was amended known as seed amendment Rules 2003.	Power of appellate authority and duty of seed analyst were slightly modified. The Seed Testing Manual was published by ICAR for reference.
1974	The Seeds [amendment] Rules 1974 were introduced.	This amendment conferred more powers on seed inspectors during crop failure.
1978	The UPOV Seed Act was introduced	The new variety should have distinctiveness, uniformity and stability for protection under this act.
1981	The Seeds [amendment] Rules 1981 were introduced	Adoption of Minimum Indian Seed Certification Standards published by Central Seed Committee was made compulsory for seed certification.
1983	The Seeds [control] Order 1983 was introduced.	This Order included Seeds as essential commodity item under the Commodity Act 1955.
1988	The New Policy on Seed Development was introduced	The new policy was formulated to provide Indian Farmers with access to the best available seeds and planting materials of domestic as well as exotic origin.
1989	Plants, Fruits and Seeds Order 1989 was introduced	This replaces the Plants, Fruits and Seeds Order 1984 and provides regulations for post entry quarantine checks.
1991	The UPOV Seed Act 1991 was introduced	The new variety should have novelty in addition to distinctiveness, uniformity and stability for protection under this modified UPOV Act.
2001	The Protection of Plant varieties and Farmers Rights Act was formulated.	A new variety with novelty, distinctiveness, uniformity and stability can be registered for protection under this Act.
2002	The National Seed Policy was formulated.	The National Seed Policy was formulated to raise Indias' share in the global seed trade by facilitating advanced scientific aspects such as biotechnology to farmers and in March 2002 first transgenic cotton was approved for commercial cultivation in India.
2003	Plant Quarantine [Regulation of Import into India] was introduced.	This Order has now replaced the Plants, Fruits and Seeds Order 1989.
2003	Protection of Plant Varieties Rules 2003 were introduced.	These rules were introduced for smooth implementation of the PPV and FR Act 2001.
2004	The New Seed Bill was formulated and introduced.	This Seed Bill has replaced the Seeds Act 1966. Registration of all seed materials is compulsory and self-certification is permitted.

Chapter 2

Indian Seed Act 1966

Q1. What is seed Act?

Ans. Seed Act refers to the legal procedures approved by the government for production and marketing of seeds. The main objective of the Seeds Act is to ensure availability of quality seeds to farmers. In India, the Seed Act was first introduced in 1966 which is called as Seeds Act 1966. This Act was amended in 1972 which is known as the Seeds amended Act 1972.

Q2. Describe in brief main featured of Seed Act of 1966.

Ans. The main features of Indian Seed Act 1966 in relation to coverage, seed category, licensing, registration, criteria for protection, period of protection, certification, compensation, farmers' privilege, penalty, and constitution of central seed committee are briefly presented below:

(1) **Coverage.** This Act has narrow coverage than seeds act of 2004. It includes agriculture and horticulture only.

(2) **Seed Category.** This act permits sale of certified seed only The name of variety, physical purity and germination percentage have to be indicated on the seed container or bag.

(3) **License.** The seed dealers, sellers and growers should have license which is compulsory.

(4) **Registration.** Registration for all kinds and varieties of seed is compulsory. No producer can grow seeds unless he is registered. It is compulsory for all cultivated varieties of field crops, vegetable crops and fruit crops. Every seed producer and dealer, and horticulture nursery has to be registered with the state government.

(5) **Criteria for Registration.** The released and notified varieties can be registered for seed multiplication.

(6) **Period of Registration.** In this seed Act the period of registration has not been defined.

(7) **Certification.** The Act does not permits self certification of seeds by accredited agencies. However, it allows the central government to recognize certification by foreign seed certification agencies. Seed producers are not permitted to self certify the performance of their seed under certain conditions.

(8) **Compensation.** The disputes about the quality of seeds have to be settled in the consumer court. Any loss in the production due to poor seed quality is claimed in the consumer court.

(9) **Farmers' Privilege.** This Act allows farmers' privilege. The Bill does not restrict the farmer's right to use or sell his farm seeds and planting material, provided he does not sell them under a brand name.

(10) **Penalty.** Any person who contravenes any provisions of the Act or imports, sells or stocks seeds deemed to be misbranded or not registered, can be punishable by a fine between Rs 100-1000/and 6 months imprisonment.

(11) **Central seed Committee.** In this Act, the central seed committee consists of Chairman [Secretary DARE], six persons related to interest in quality seeds, two representatives of seed growers and one representative from each state. Members are nominated for a period of two years and can be re-nominated for a similar period.

Q3. What are merits of Seed Act 1966?

Ans. There are several merits of seed act 1966. Merits of this act related to biodiversity, seed monopoly, tracing defective seed lot, certification, farmers rights, seed quality, duration of registration and effects on seed industry and productivity are briefly presented as follows:

(1) **Effect on Biodiversity:** This seed act permits cultivation of both new varieties as well as land races. Thus it does not t have any adverse effect on crop biodiversity. Thus there is no Thus there is no danger of uniformity.

(2) **No Monopoly of Seed:** When the seed act 1966 was approved, there were no multinational seed companies as a result there was no risk of monopoly of seed. The entire seed production was under government organizations.

(3) **Easy to Trace Defective Seed Lot:** The Seed Act does provides for a mechanism to trace back a packet of seed to the dealer, processor and producer. Thus tracing of defective lot was easy to rectify any deficiencies in the supply chain.

(4) **No Self Certification:** There was no provision of self certification in the Seed Act 1966. The certification is done by government seed certification agencies. Thus it does not have any scope for false certification of the seed.

(5) **Permits Farmers' Rights.** It permits farmers' rights. The farmers need not to register their varieties.

TABLE 2-1: Merits and Demerits of Seed Act 1966

Sl.No.	Particulars	Comments or Explanation
A	**Merits**	
1	Farmers privilege	Farmers' rights or privilege are protected.
2	Effect on biodiversity	There is no adverse effect on biodiversity because this act permits cultivation of both improved varieties and land races.
3	Certification	Certification is permitted by government organizations only which ensures quality of seed.
4	Tracing defective seed lot.	The tracing of defective seed lot is easy.
5	Registration	It permits registration till the variety is in seed production chain.
6	Licensing	Licensing is compulsory for seed sellers which ensures seed quality.
B	**Demerits**	
1.	Coverage	Coverage is narrow. It includes Agriculture and horticulture only.
2	Central Seed Committee	Central Seed Committee does not have representative of private seed industry.
3	Compensation	Not clearly defined.
4	Seed Sale	Not clearly defined
5	Benefit Sharing	There is no provision of benefit sharing.
6	Period of protection	Not defined.

(6) **Ensures Quality Seed**. This act ensures availability of good quality seeds to farmers by imposing penalties on the sale of spurious seeds.

(7) **Registration for long Duration**. This act provides registration for a long period. Because the period of registration has not been specified.

(8) **Promotes Seed Industry**. The introduction of Seed Act, 1966 resulted in the fast development of seed industry in India due to involvement of representatives of multinational seed companies or industries in the central seed committee.

(9) **Increase in Productivity**. The Seed Act has helped in increasing crop productivity due to availability of better quality seeds.

Q4. What are demerits of Seed Act 1966?

Ans. There are some demerits of Seed Act 1966. Demerits related to coverage, constitution of central seed committee, *etc.* are presented below:

(1) **Narrow Coverage**. This act has narrow coverage [agriculture and horticulture] than seed act 2004 which covers, forestry, plantation crops, medicinal and aromatic plants, in addition to agriculture and horticulture.

(2) **Central Seed Committee**. In the Seed Act 1966, the size of central seed committee is bigger than that of Seed Act 2004. There was no representative from the private seed industry.

(3) Compensation: It is not clear whether the compensation would include the value of the crop or only the cost of the seed.

(4) Seed Sale: It is not clear whether a seed producer may sell seed which is registered by a different producer.

(5) Benefit Sharing: The Act does not have the provision of benefit sharing unlike the Convention on Biological Diversity and the PPVFR Act.

Chapter 3

The Seed Rules, 1968

Q1. What are seed Rules 1968?

Ans. The Seeds Rules, 1968 under Seed Act, 1966 (Act No. 54 of 1966) were formulated and approved for implementation of Seed Act, 1966.These rules consist of 11 Parts and 39 sections.

Q2. Explain briefly Seed Rules 1968 on the basis of scope and functions of central seed committee.

Ans. Salient features of seed rules, 1968 related to scope and functions of central seed committee,are briefly presented as follows:

(1) Scope and Coverage: These rules may be called the Seeds Rules, 1968. and extend to whole of India. These rules cover agriculture and horticulture. Various terms used related to seed rules have also been defined [Part 1].

(2) Central Seed Committee: This committee has several functions. Main functions of this committee are [Part 2]:

   (i) To recommend the rate of fees to be levied for analysis of samples by the Central and States Seed Testing Laboratories and for certification by the certification agencies;

   (ii) To advice the Central or State Governments on the suitability of seed testing laboratories;

   (iii) To send its recommendations and other concerning records to the Central Government;

   (iv) To recommend the procedure and standards for certification, tests and analysis of seeds; and

   (v) To carry out any other functions related to these rules.

Q3.   **What are the functions of Central Seed Laboratory?**

Ans.   In addition to the functions entrusted to the/central Seed Laboratory by the Act, the Laboratory shall carry out the following functions [Part 3].

    (i) Initiate testing programs in collaboration with the State Seed Laboratories designed to promote uniformity in test results between all seed laboratories in India;

    (ii) Collect data continually on the quality of seeds found in the market and make this data available to the Committee; and

    (iii) Carry out such other functions as may be assigned to it by the Central Government from time to time.

Q4.   **What are the functions of Seed Certification Agency?**

Ans.   Main functions entrusted to the certification agency by the Act are as follows:

    (i) Certify seeds of any notified kinds or variety;

    (ii) Outline the procedure of submission of applications and for growing, harvesting, processing, storage and labeling of seeds intended for certification till the end to ensure that seed lots finally approved for certification are true to variety and meet prescribed standards for certification under the Act or these rules;

    (iii) Maintain a list of recognized breeders of seeds;

    (iv) Verify, upon receipt of an application, for certification, that the variety is eligible for certification that the seed source need for planting was authenticated and the record of purchase is in accordance with these rules and the fees has been paid;

    (v) Take sample and inspect seed lots produced under the procedures laid down by the certification agency and have such samples tested to ensure that the seed conforms to the prescribed standards of certification;

    (vi) Inspect seed processing plants to see that the admixtures of other kinds and varieties are not introduced.

    (vii) Ensure that action at all stages *e.g.* field inspection, seed processing, plant inspection, analysis of samples taken and issue of certificates (including tags, marks, labels and seals) is taken expeditiously;

    (viii) Carry out educational programs designed to promote the use of certified seed including a publication listing certified seed growers and sources of certified seed;

    (ix) Grant certificate (including tags, marks, labels and seals *etc.*) in accordance with the provisions of the Act and these Rules;

    (x) Maintain such records as may be necessary to verify that seed plants for the production of certified seed were eligible for such planting under these rules;

(xi) Inspect fields to ensure that the minimum standards for isolation, rouging (where applicable) use of male sterility (where applicable) and similar factors are maintained at all times, as well as ensure that seed borne diseases are not present in the field to a greater than those provided in the standards for certification.

Q5. Describe Marketing or labeling of certified seed.

Ans. Marketing and labeling decides about various aspects such as labeling, contents of label and procedure of labeling as discussed below:

(a) Responsibility for marking or labeling: When seed of a notified kind or variety is offered for sale, each container shall be marked or labeled in the manner hereinafter specified. The person whose name appears on the mark or label shall be responsible for the accuracy of the information required to appear on the mark or label so long as seed is contained in the unopened original container.

Provided, however, that such person shall not be responsible for the accuracy of the statement appearing on the mark or label if the seed is removed from the original unopened container, or he shall not be responsible for the accuracy of the germination statement beyond the date of validity indicated on the mark or label.

(b) Contents of the mark or label: There shall be specified on every mark or label

   i. Particulars, as specified by the Central Government under Clause (b) of section 6 of the act.

   ii. A correct statement of the net content in terms of weight and expressed in metric system.

   iii. Date of testing.

   iv. If the seed in container has been treated.

     a. Statement indicating that the seed has been treated.

     b. The commonly accepted chemical of abbreviated chemical (generic) name of the applied substance; and

     c. If the substance of the chemical used for treatment, and present with the seed is harmful to human beings or other vertebrate animals, a caution statement such as "Do not use for food, feed or oil purposes". The caution for mercurial and similarly toxic substance shall be the word "Poison" which shall be in type size, prominently displayed on the label in red.

   v. The name and address of the person who offers for sale, sells or otherwise supplies the seed and who is responsible for its quality;

   vi. The name of the seed as notified under section 5 of the act.

(c) Manner of marking or labeling the container under clause (c) of section 7 and clause (b) of section 17:

(i) The mark or label containing the particulars of the seed shall appear on each container of seed or on a tag or mark or label attached to the container in a conspicuous place on the inner most container in which the seed is packed and on every other covering in which that container is packed and shall be legible.

(ii) Any transparent cover or any wrapper, case or other covering used solely for the purpose of packing of transport or delivery need not be marked or labeled.

(iii) Particulars that are required to be displayed on a label on the container, be attached, painted or otherwise indelibly marked on the container.

(d) Mark or label not to contain false or misleading statement:

The mark or label shall not contain any statement, claim, design, device, fancy name or abbreviation which is false or misleading in any particular concerning the seed contained in the container.

(e) The mark or label shall not contain any reference to the Act, or any of these rules or any comment on, or reference to, or explanation of any particulars or declaration required by the Act or any of these rules which directly or by implication contradicts, qualifies or modifies such particulars or declaration.

(f) Denial of responsibility for mark or label content prohibited:

Nothing shall appear on the mark or label or in any advertisement pertaining to any seed of any notified kind or variety which shall deny responsibility for the statement required by or under the Act to appear on such mark, label or advertisement.

Q6. What are requirements to carry out the seed business?

Ans. There are certain conditions that are required for a person to carry out the seed business. Important requirements to carry out seed business are as follows [Part 6].

(i) No person shall sell, keep for sale, offer to sell barter or otherwise supply any seed of any notified kind or variety, after the date recorded on the container, mark or label as the date up to which the seed may be expected to retain the germination not less than that prescribed standard.

(ii) No person shall alter, obliterate or deface any mark or label attached to the container of any seed.

(iii) Every person selling, keeping for sale, offering to sell, bartering or otherwise supplying any seed of notified kind or variety [under Section 7], shall keep over a period of three years a complete record of each lots of seed sold except that any seed sample may be discarded one year after the entire lot represented by such sample has been disposed off. The sample of seed kept as part of the complete record shall be as large as the size notified in the official Gazette.This sample, if required to be tested only for determining the purity.

(iv) Classes and sources of certified seed:

There shall be three classes of certified seed, namely, foundation, registered and certified, and each class shall meet the following standards for that class.

a. Foundation seed: This shall be the progeny of breeder's seed, or be produced from foundation seed which can be clearly traced to breeder's seed. Its production shall be supervised and approved by a seed certification agency in such a way to maintain specific genetic purity and meet certification standards prescribed for the crop under certification.

b. Registered seed: This shall be the progeny of foundation seed and is produced in such a way to maintain its genetic purity and meet germination standards for the particular crop being certified.

c. Certified seed: This shall be the progeny of registered or foundation seed ans is produced in such a way to maintain genetic purity and germination standards specified for the particular crop being certified.

At the discretion of the certification agency (when considered necessary to maintain adequate seed supplies) certified seed may be progeny of certified seed provided this reproduction may not exceed three generations and provided further that it is determined by the seed certification agency that the genetic purity will not be significantly altered.

Q7. Describe briefly the procedure of Certification of Seeds.

The application, fees and certificate are explained as follows:

(i) **Application for the grant of a certificate:** Every application for the grant of a certificate shall be made in prescribed Form and contain the following particulars,

a. The name, profession and place of residence of the applicant;

b. The name of the seed to be certified, its notified kind or variety;

c. Class of the seed;

d. Source of the seed;

e. Limits of germination and purity of the seed;

f. Mark or label of the seed

(ii) **Fees:** Every application shall be accompanied by a fee of Rs,25 in cash.

(iii) **Certificate:** Every certificate granted shall be in Form II and shall be granted by the certification agency after making enquiries and satisfying itself.

I. The person to whom the certificate is granted shall attach a certification tag to every container of the certified seed and shall follow the provisions in respect of marking or labeling provided by or under the act.

II. The certification tag shall contain the following particulars namely.

a. Name and address of the certification agency.

b. Kind and variety of the seed.

c. Lot no or other mark of the seed.

d. Name and address of the certified seed producer.

e. Date of issue of the certificate and of its validity.

f. An appropriate sign to designate certified seed.

g. An appropriate word denoting the class designation of the seed.

III. The color of the certification tag shall be white for foundation seed purple for registered seed and blue for certified seed.

IV. The container of the certified seed shall carry a seal of such material and in form as the certification agency may determine and no container carrying a certification tag shall be sold by the person if the tag or seal has either been tampered with or removed.

V. The certification tag on the container shall specify.

a. The period during which the seed shall be used for sowing or planting.

b. That the use of seed after the expiry of the validity period by any person is entirely at his risk and the holder of the certificate shall not be responsible for any damage to the buyer of the seed.

c. That no one should purchase the seed if the seal or the certification tag has been tampered with.

VI. The holder of the certificate shall keep record of the details of each lot of the seed which is issued for sale in such form as to be available for inspection and to be easily identified by reference to the number of the lot as shown in the certification tag of each container and such records shall be retained in the case of a seed for which expiry date is fixed for a period of two years from the expiry of such date.

VII. The holder of the certificate shall allow any Seed Inspector, authorized in writing by the certification agency in that behalf, to enter with or without prior notice, the premises where the seeds are grown, processed and sold and to inspect premises, plant and the process of processing at all reasonable hours.

VIII. The holder of the certificate shall allow the Seed Inspector, authorized in writing by the certification agency, to inspect all registers and records maintained under these rules and to take samples of the seeds and shall supply to the Seed Inspector such information as he may require for the purposes of ascertaining whether the conditions subject tom which the certificate has been granted, have been complied with.

IX. The holder of the certificate shall on request furnish to the certification agency from every lot of the seed or from such lot or lots as the said agency may from time to time specify, a sample of such quantity as the agency may consider adequate for any examination required to be made.

X.   If the certificati0on agency so directs, the holder of the certificate shall not sell or offer for sale any lot in respect of which a sample is furnished under the proceeding clause until the agency authorizes the sale of such lot.

XI.   The holder the certificate shall, on being directed by the certification agency that any part of a lot has been found by the said agency not to conform to prescribed standards of quality or purity specified by or under the Act, withdraw the remainder of that lot from sale and so far as may, in the particular circumstances of the case, be practicable, recall all issues already made from that lot.

XII.   The holder of the certificate shall comply with the provisions of the Act and these Rules and with the directions given after not less than one month's notice by the certification agency to such holder.

The Certification agency shall, before granting the certificate, ensure that the Seed conforms to the standards laid down in the Manual known as "Indian Minimum Seed Certification Standards" published by the Central Seed Committee, as amended from time to time".

Q8. How Appeals are made as per seed Rules 1968?

Ans.   All appeals are made in a prescribed form and a prescribed fee has to be paid as discussed below [Part 8].

(i)   Every memorandum of appeal shall be in writing and shall be accompanied by a copy of the decision of the certification agency against which it has been preferred and shall set forth concisely and under distinct heads the grounds of objections to such decision without any argument, or narrative.

(ii)   Every such memorandum of Appeal shall be accompanied by a treasure receipt for sum of 100/- rupees.

(iii)   Every such memorandum of appeal may be presented either in person or through an agent duly authorized in writing in this behalf by the appellant or may be sent by the registered post.

(iv)   In deciding appeals under the Act the Appellate authority shall exercise all the powers which a Court has and shall follow the same procedure with a Court follows in deciding appeals from the decree or order of an original Court under the Code of Civil Procedure, 1908 (5 of 1908).

Q9. What are Qualifications and Duties of Seed Analysts?

Ans.   The qualifications and duties of seed analysts are given as follows:

(1) Qualifications of Seed Analysts: A person shall not be qualified for appointment as Seed Analyst unless he.

i. Possesses a Master's or equivalent degree in Agriculture or Agronomy or Botany or Horticulture of a University recognized for this purpose by the Government and has had not less than one year's experience in seed technology; or

ii. Possessed a Bachelor's degree in Agriculture or Botany of a University recognized for this purpose by the Government and has had not less than three years' experience in seed technology.

(2) Duties of a seed analyst:

(i) On receipt of a sample for analysis the Seed Analyst shall first ascertain that the mark and the seal are intact.

(ii) The Seed analyst shall analyze the samples in accordance with the procedures laid down in the Seed Testing Manual published by the Indian Council of Agriculture Research.

(iii) The Seed Analyst shall deliver in Form VII, a copy of report of the result of the seed not later than 30 days from the date of receipt of samples sent by the Seed Inspector.

(iv) The Seed Analyst shall from time to time forward to the State Government the reports giving the result of analytical work done by him.

Q10. What are Qualifications and Duties of Seed Inspectors?

Ans. The qualifications and duties of seed inspectors are given as follows:

(1) **Qualifications:** A person shall not be qualified for appointment as Seed Inspector unless he is a graduate in Agriculture of a University recognized for the purpose by the Government and has had not less than one year's experience in seed production, or seed development in seed analysis or testing in seed testing laboratory.

(2) **Duties of a Seed Inspector:** In addition to the duties specified by the Act, the Seed Inspector shall a. Inspect as frequently as may be required by certification agency all places used for growing storage or sale of any seed of any notified kind or variety;

(i) Satisfy himself that the conditions of the certificates are being observed;

(ii) Procedure and send for analysis, if necessary, samples of any seeds, which he has reason to suspect are being produced, stocked or sold or exhibited for sale on contravention of the provisions of the Act or these rules;

(iii) Investigate any complaint, which may be made to him in writing in respect of any contravention of the provisions of the Act or these rules.

(iv) Maintain a record of all inspections made and action taken by him in the performance of his duties including the taking of samples and the seizure of stocks and submit copies of such record to the Director of Agriculture or the certification agency as may be directed in this behalf.

(v) When so authorized by the State Government detain imported containers which he has reason to suspect contain seeds, import of which is prohibited except and in accordance with the provisions of the Act or these rules.

(vi) Institute prosecutions in respect of breaches of the Act or these rules.

(vii) Perform such other duties as may be entrusted to him by the State Government.

Q11. What are the actions to be taken by the Seed Inspector if a complaint is lodged with him?

Ans. The Seed Inspector can take following actions if a complaint is lodged with him.

(1) If farmer has lodged a complaint in writing that the failure of the crop is due to the defective quality of seeds of any notified kind or variety supplied to him, the Seed Inspector shall take in his possession the marks or labels, the seed containers and a sample of unused seeds to the extent possible from the complaint for establishing the sources of supply of seeds and shall investigate the causes of the failure of his crop by sending samples of the lot to the Seed Analyst for detailed analysis at the State Seed Testing Laboratory. He shall thereupon submit the report of his findings as soon as possible to the competent authority.

(2) In case, the Seed Inspector comes to the conclusion that the failure of the crop is due to the quality of seeds supplied to the farmer being less than the minimum standards notified by the Central Government, he shall launch proceedings against the supplier for contravention of the provisions of the Act or these Rules

Q12. Describe in brief the procedure of Sealing, fastening, dispatch and analysis of samples.

Ans. The procedure of sampling, packing, fastening and sealing samples, containers to be used, labeling, form of seed report, fees, *etc.* are briefly explained as follows:

(i) **Manner of taking samples:** Samples of any seed of any notified kind of variety for the purpose of analysis shall be taken in a clean dry container which shall be closed sufficiently tight to prevent leakage and entrance of moisture and shall be carefully sealed.

(ii) **Containers to be labeled and addressed:** All containers containing samples for analysis shall be properly labeled and the parcels shall be properly addressed. The label on any sample of seed sent for analysis shall bear.

    a. Serial number;

    b. Name of the sender with official designation, if any;

    c. Name of the person from whom the sample has been taken.

    d. Date and place of taking the sample;

    e. Kind or variety of the seed for analysis;

    f. Nature and quantity of preservative, if any, added to the sample.

(iii) **Manner of packing, fastening and sealing the samples:** Samples of seed sent for analysis shall be packed, fastened and sealed in the following manner.

    a. The stopper shall first be securely fastened to as to prevent leakage of the containers in transit;

    b. The container shall then be completely wrapped in fairly strong thick paper. The ends of the paper shall be neatly folded in and affixed by means of gum or other adhesive.

    c. The paper cover shall be further secured by means of strong twine or thread both above and across the container, and the twine or thread shall then be fastened on the paper cover by means of sealing wax on which there shall be test four distinct and clear impression of the seal of the sender of which one shall be at the top of the packet, one at the bottom and the other two on the body of the packet. The knots of the twine or thread shall be covered by means of sealing wax bearing the impression of the seal of the sender.

(iv) **Form of order:** The order to be given in writing by the Seed Inspector shall be in Form III

(v) **Form of receipt of records:** When a Seed Inspector seizes any record, register, document or any other material he shall issue a receipt in Form IV to the person concerned.

(vi) **Samples how to sent to the Seed Analyst:** The container of sample for analysis shall be sent to the Seed Analyst by registered post or by hand in a sealed packet enclosed together with a memorandum in Form V in an outer cover addressed to the Seed Analyst.

(vii) **Memorandum and impression of seal to be sent separately:** A copy of the memorandum and a specimen impression of the seal used to seal the packet shall be sent to the Seed Analyst separately by registered post or delivered to him or to any person authorized by him.

(viii) **Addition of preservatives to samples:** Any person taking a sample of seed for the purpose of analysis under the Act may add a preservative as may be specified from time to time to the sample the nature and quantity of the preservative added shall be clearly noted on the label to be affixed to the container.

(ix) **Nature and quantity of the preservative to be noted on the label:** Whenever any preservative is added to a sample the nature and quantity of the preservative added shall be clearly noted on the label to be affixed to the container.

(x) Analysis of the sample: On receipt of the packet, it shall be opened either by the Seed Analyst or by an officer authorized in writing in that behalf by the Seed Analyst, who shall record the condition of the seal on the packet. Analysis of the sample shall be carried out at the State Seed Laboratory in accordance with the procedure laid down by the Central Government.

(xi) Form of notice: The notice to be given to the person from whom the Seed Inspector intends to take sample shall be in Form VI.

(xii) Form of report: The report of the result of the analysis shall be delivered or sent in Form VII.

(xiii) Fees: The fees payable in respect of the report from the Central Seed Laboratory shall be Rs.10/- per sample of the seed analyzed.

(xiv) Retaining of the sample: The sample of any seed shall be retained under a cool, dry environment to eliminate the loss of viability and in insect proof or rat proof containers. The containers shall be dusted with suitable insecticides and the storage room fumigated to avoid infestation of samples by insects. The samples shall be packed in good quality containers of uniform shape and size before storage.

(1) A white tag shall be used for foundation seed.

(2) A purple tag shall be used for registered seed.

(3) A blue tag shall be used for certified seed.

(4) Certification shall be valid for the period indicated on the tag provided seed is stored in cool dry environment.

Chapter 4

Seed Amendment Rules

Introduction

In India, the Seed Act was first introduced in 1966 which is called as Seeds Act 1966. This Act was amended in 1972, 1973, 1974 and 1981. A brief discussion of these amendments is presented as follows:

Q1. Explain briefly seeds amendment Act, 1972

Ans. The first amendment to the seed act 1966 was approved by the central government on 9th September, 1972. This is called the seed amendment Act, 1972. In this act three sections (Section 2, 9 and 25) were amended and one new section (Section 8 A to E) was added. These are explained as follows;

(i) In section 2 (definitions), Jute seeds were inserted after seeds of cattle fodder.

(ii) In section 9 (grant of certificate by certification agency), it was inserted that the seed should meet prescribed standard of germination and purity (instead of minimum limit of germination and purity as given in Seed Act, 1966).

(iii) In section 25 (power to make rules), it should be inserted "the standard to which seed should conform (sub-section 2 after clause f) and the word two successive seasons should be replaced by two or more successive seasons.

(iv) Insertion of new section (Section 8 A to E): This section has provision for establishment of Central seed Certification Board which will advise central government on all matters of seed certification and coordinate the functioning of seed certification agencies. The board will consist of a Chairman (to be nominated by central government), four members from department of agriculture (Not less than director of agriculture), three members from agriculture universities (not less than director of research), ten members related to seeds and three members as representatives of seed producer. The membership is for a period of two years. The Central Government shall appoint a person as the Secretary of the Board.

Q2. Explain briefly seeds amendment Rules, 1973

Ans. These amendment rules were approved by the government of India, Ministry of Agriculture (Department of Agriculture) on 30th June, 1973 and may be called Seed Amendment Rules, 1973 of Seed Rules, 1968. The important amendments are as follows:

(1) In rule 19 of the Seeds Rule, 1968 (under appeal), the judicial powers of authority have been omitted.

(2) In rule 21 of the said rules for sub-rules (2) and (3) the following sub-rules shall be substituted, namely:

(i) The Seed Analyst shall analyze the samples in accordance with the procedures laid down in the Seed Testing Manual published by the Indian Council of Agricultural Research as amended from time to time."

(ii) The Seed Analyst shall deliver in Form VII, a copy of the report of the result of analysis to the persons specified in sub-section (1) of Section 16, as soon as may be but not later than 30 days from the date of receipt of samples sent by the Seed Inspector under sub-section (2) of the Section 15".

(3) In rule 23 of the said rules, in clause (h) for the words competent authority "the words" State Government shall be substituted.

Q3. Discuss in brief seeds amendment Rules, 1974

Ans. These amendment rules were approved by the government of India, Ministry of Agriculture (Department of Agriculture) on 29th April, 1975. These Rules may be called Seed Amendment Rules, 1974 of Seed Rules, 1968. The important amendments are as follows:

After rule 23 of the said rules, the following rule shall be inserted namely "23-A. Action to be taken by the Seed Inspector if a complaint is lodged with him.

(1) If farmer has lodged a complaint in writing that the failure of the crop is due to the defective quality of seeds of any notified kind or variety supplied to him, the Seed Inspector shall take in his possession the marks or labels, the seed containers and a sample of unused seeds to the extent possible from the complaint for establishing the source of supply of seeds and shall investigate the causes of the failure of his crop by sending samples of the lot to the Seed Analyst for detailed analysis at the State Seed Testing Laboratory. He shall thereupon submit the report of his findings as soon as possible to the competent authority.

(2) In case, the Seed Inspector comes to the conclusion that the failure of the crop is due to the quality of seeds supplied to the farmer being less than the minimum standards notified by the Central Government, can launch proceedings against the supplier for contravention of the provisions of the Act or these Rules."

Corrigendum [No. 7-15/74-SD New Delhi, dated the 30th/31st Jan., 1976]

In the Seeds (Amendment) Rules, 1974, published with the notification of the Government of India in the Ministry of Agriculture and Irrigation (Department of Agriculture) (No.GSR 211 (E), dated the 29th April, 1975, in the Gazette of India Extraordinary, Part II, Section 3, Sub-section (i), dated the 29th April, 1975:- at page 863

(*i*) in line 3 for "to" read "the";

(*ii*) in line 11, for "makes" read "marks";

(*iii*) in line 20 and 21, for "the Central Government launch proceedings", read "the Central Government, shall launch proceedings".

Q4. Give a brief account of seeds amendment Rules, 1981

Ans. These amendment rules were approved by the government of India, Ministry of Agriculture (Department of Agriculture) on 10th June, 1981. These Rules may be called the Seed Amendment Rules, 1981 of Seed Rules, 1968. The important amendments are as follows:

(1) These rules may be called the Seeds (Amendment) Rules, 1981.

(2) They shall come into force on the date of their publication in the Official Gazette. After rule 17 of the Seeds Rules, 1968, the following rule shall be inserted, namely:

17-A The Certification agency shall, before granting the certificate, ensure that the seed conforms to the standards laid down in the Manual known as "Indian Minimum Seed Certification Standards" published by the Central Seed Committee, as amended from time to time."

Q5. Describe in brief main features of seeds bill, 2004

Ans. In March, the Cabinet had approved the Seeds Bill, 2004 that seeks to regulate the quality of hybrid seeds and check the sale of spurious seeds in the country, besides increasing private participation in seed production and distribution. Amendments to Seed Bill 2004 were approved by the central government on 20th October, 2010.

(1) The additional amendments provided for submission of seed related periodic returns to state governments and enhancement in penalties of offences.

(2) The government has also included a provision to nominate chairperson of Protection of Plant Varieties and Farmers Rights Authority and National Biodiversity Authority to the Central Seed Committee.

(3) The Bill seeks to repeal and replace existing Seeds Act, 1966, for it does not deal with the quality control of GM seeds, as they are generally not notified.

(4) The legislation seeks to regulate the quality of seeds and planting materials, curb the sale of spurious and poor quality seeds, increase private participation in seed production and distribution, and liberalize imports of seeds.

(5) Provisions of labeling, seed health, expected performance and compensation to farmers have been included to ensure public accountability.

(6) Innovations include compulsory registration, enabling government to exclude certain varieties of seeds on ground of public health, environment, provision for expected performance, seed health and farmer's compensation, *etc.*

Q6. Give a brief account of various amendments of Seed Act, 1966.

Ans. A brief account of various amendments of Seed Act, 1966 is presented in tabular form as follows:

TABLE 4-1: Summary of Seed Amendments

Sl.No.	Seed Amendment	Major Amendments
1	Seeds (Amendment) Act, 1972	(i) Jute seeds were included in seed Act, 1966. (ii) Seed should meet prescribed standards of germination and purity. (iii) More than two successive seasons. (iv) Provision for establishment of Central seed certification board was added.
2	Seeds (Amendment) Rules, 1973	(i) Judicial powers of authority have been omitted. (ii) Samples to be analyzed as per seed testing manual of ICAR. (iii) Testing report to be submitted in not more than 30 days. (iv) The word competent authority is replaced by State government
3	Seeds (Amendment) Rules, 1974	If there is crop failure due to poor seed quality, Inspector has to take action against seed supplier on complaint of farmers.
4	Seeds (Amendment) Rules, 1981	The Certification agency shall, before granting the certificate, ensure that the seed conforms to the prescribed standards.
5	Amendments to Seeds Bill, 2004	Chairpersons of PPV and FR Act and Biodiversity Act were included in the Central Seed Committee.

Seeds (Control) Order, 1983

Q1. What is Seeds Control Order, 1983?

Ans. The Seeds (Control) order, 1983 under Seed Act, 1966 (Act No. 54 of 1966) was approved by the government of India, Ministry of Agriculture (Department of Agriculture and Cooperation) on 30th December, 1983 and may be called Seed control Order, 1983. This order consists of four parts. Part 1 deals with title, scope and definitions of various terms. Part 2 deals with licensing of dealers, part 3 deals with appointment of licensing authority, appointment of inspectors, inspection and punishment, time limit for analysis, suspension/cancellation of license and appeals. Part 4 deals with amendment of license, maintenance of records and submission of returns, *etc*. These are briefly presented as follows:

Q2. What is the scope of Seeds Control Order, 1983?

Ans. This order may be called the Seeds (Control) Order, 1983. It extends to the whole of India. It shall come into force on the 30th December of 1983. The central Government hereby declare the following seeds used for sowing or planting (including seedlings and tuber, bulbs, rhizomes, roots, cuttings and all types of graft and other vegetatively propagated material of food crops or cattle fodder) to be essential commodities for the purpose of the said Act, namely.

(i) Seeds of food crops and seeds of fruit and vegetables.

(ii) Seeds of cattle fodder and

(iii) Jute seeds.

Q3. Explain various aspects of licensing for Seeds Control Order, 1983

Ans. According to this seed order, it is compulsory for seed dealers to obtain license to carryout seed business. No person shall carry on the business of selling, exporting or importing seeds at any place except under and in accordance with the terms and conditions of license granted to him under this Order. Various aspects of licensing are briefly presented as follows:

Sl.No.	Particulars	Brief explanation
1	Scope and Coverage	This seeds order extents to whole of India and covers seeds of food crops, fruits and vegetable crops, fodder crops and jute.
2	Compulsory Licensing	This section deals with various aspects of licensing such as grant or refusal of license, renewal of license, period of validity of license, *etc.*
3	Licensing Authority	This section deals with appointment of licensing authority, seed inspectors, powers of seed inspectors, time limit for analysis report, cancellation or suspension of license, appeal *etc.*
4	Amendment of License	This section deals with amendment of license, maintenance o0f records and submission of monthly returns.

Q4. Explain procedure of obtaining license for Seeds Control Order, 1983

Ans. The procedure of obtaining license for Seeds Control Order, 1983 is presented as follows:

(i) **Application for License:** Every person desiring to obtain a license for selling, exporting or importing of seeds shall make an application in duplicate in Form "A" together with a fee of Rs.50/- for license to licensing authority.

(ii) **Grant and Refusal of License:** The licensing authority may, after making such enquiry as it thinks fit, grant a license in Form "B" to any person who applies for it. However, issue of license to a person can be refused under following conditions.

    a. If his earlier license granted under this Order is under suspension. The refusal would for the period of such suspension.

    b. If his earlier license granted under this Order has been cancelled. The refusal would for a period of one year from the date of cancellation.

    c. If a person has been convicted under the Essential Commodities Act, 1955(10 of 1955) or any order issued there under within three years preceding the date of application.

When the licensing authority refused to grant license to a person who applied for it, he shall record his reasons for doing so.

(iii) **Period of validity of license:** Every license under this order, shall, unless previously suspended or cancelled, remain valid for three years from the date of its issue.

(iv) **Renewal of License:** For renewal of license, the license holder has to give an application in the prescribed form (Form C) to the licensing authority before expiry of the license. The fee of rupees twenty is to be paid for renewal. If any application for renewal is not made before the expiry of the license, but is made within one month from the date of expiry of the license, the license may be renewed on payment of additional fee of rupees twenty-five, in addition to the fee for renewal of license.

(v) Display of Stock and Price List: Every seed dealer has to daily display the opening and closing stocks, of different seeds held by him and a price list of different seeds available with him.

(vi) Issue of Cash/Credit Memo to Purchaser: Every dealer shall give a cash or credit memorandum to a purchaser of seeds.

Q5. Describe briefly licensing Authority for Seeds Control Order, 1983

Ans. For effective implementation of the order, the State Government has to appoint licensing authority and inspectors areas wise and delegate them certain powers. The power of inspection and punishment are also decided. A seed inspector has the following powers.

(i) He can ask any dealer to give information about purchase, storage and sale of seeds by him;

(ii) He can enter and search any premises where any seed is stored or exhibited for sale to ensure compliance with the provisions of this Order;

(iii) He can draw samples of seeds meant for sale, export and seeds imported, and send the same to a laboratory to ensure that the sample conforms to standard of quality claimed.

(iv) He can seize or detain any seed in respect of which he has reason to believe that contravention of this order has been committed or is being committed,

(v) He can seize any books of accounts or document relating to any seed in respect of which he has reason to believe that a contravention of this Order has been committed and is being committed. Every seed dealer will have to cooperate with the seed inspector in this matter.

Q6. What are Duties of Inspector as per Seeds Control Order, 1983?

Ans. The Inspector shall give receipt, in respect of the books of accounts or documents seized, to the person from whom they have been seized.

The seized books of account or documents shall be returned to the person from whom the same had been sized after copies thereof or extracts there from as certified by such person have be taken.

When any seed is seized by an Inspector, he shall forthwith report the fact of such seizure to a Magistrate in his jurisdiction for further action.

Q7. What is the Time Limit for seed sample analysis as per Seeds Control Order, 1983?

Ans. The laboratory to which a sample has been sent by an Inspector for analysis under this order shall analyze the said samples and send the analysis report to the concerned Inspector within 60 days from the date of receipt of the sample in the laboratory.

Q8. Describe briefly suspension/cancellation and amendment of license as per Seeds Control Order, 1983

Ans. A brief account of suspension/cancellation and amendment of license as per Seeds Control Order, 1983 is presented as follows:

(i) **Suspension/Cancellation of license:** If the license has been obtained by misrepresentation or any provision of the order/condition of the license has been contravened, the licensing authority may suspend or cancel the license. However, the license holder is given an opportunity to clarify before such action.

(ii) **Amendment of license:** The licensing authority may, on receipt of a request in writing together with a fee of rupees ten from a dealer, amend the license of such dealer.

(iii) **Appeal:** If any person has been refused to grant, amend or renew the license for sale, export or import of seeds or his license has been cancelled/suspended, he may appeal within sixty days from the date of the order, to such authority as the State Government may specify in this behalf, and the decision of such authority shall be final. The appeal application shall accompany an appeal fee of rupees fifty.

Q9. Write short note on maintenance of records and submission of returns as per Seeds Control Order, 1983

Ans. Every dealer shall maintain such books, accounts and record relating to his business as may be directed by the State Government. Moreover, every dealer shall submit monthly return relating to his business for the preceding month in Form "D" to the licensing authority by the 5th day of every month.

Chapter 6

New Seed Policy, 1988

Q1. What are the objectives of New Seed Policy, 1988?

Ans. The New Seed Policy was approved by the Government of India, Ministry of Agriculture [Department of agriculture and Cooperation] on 16th September, 1988. This seed policy has been approved with the following objectives.

(i) To provide the best planting materials to the farmers available in the world.

(ii) To increase agriculture productivity and thereby increasing farm income.

(iii) To increase export earnings through planting material.

Q2. What are the materials covered by New Seed Policy, 1988?

Ans. The New Seed Policy covers the import of following plant materials:

(i) Seeds of wheat and paddy.

(ii) seeds of coarse cereals/pulses/oil seeds;

(iii) Seeds of vegetable and flower;

(iv) Bulbs/tubers of flowers;

(v) cuttings/saplings/bud/wood *etc.*, of flowers; and

(vi) Seeds and planting material of fruits.

Q3. What are guidelines for import of seeds of coarse cereals/pulses/oil seeds?

Ans. It is necessary to increase the production of coarse cereals, oilseeds and pulses by making available high yielding seeds to the farmers. At the same time, repetitive bulk imports of these seeds cannot be permitted indefinitely. Based on these considerations, the new import policy for seeds for sowing of coarse cereals, oilseeds and pulses is detailed in the following paragraphs. Agreement to supply parent line seeds/breeder seeds/technology NSC, SSC's may import Issue of Import License by the CCI and E Seed for trial evaluation by ICAR Recommendation of DAC Publicity through Media

(i) **Period of Import:** The import of seeds of coarse cereals, pulses and oilseeds for sowing may be allowed for a period not exceeding two years by companies, which have technical/financial collaboration agreements for production of seeds with companies abroad, provided the foreign supplier agrees to supply parent line seeds/nucleus or breeder seeds/ technology to the Indian company within a period of two years from the date of import of the first commercial consignment after its import has been recommended by DAC.

(ii) **Organization to Import:** The National Seeds Corporation and the State Seeds Corporation may be permitted to import of coarse cereals, oilseeds and pulses on the recommendations of DAC.

(iii) **Grant of Import License:** Plant Protection Advisor (PPA) or any other officer notified for the said purpose issues the import permit based on the recommendation of DAC. DAC may, within 30 days of the receipt of the application, reject the application or recommend it to the Central Committee for Import and Export (CCI and E) for grant of an import license to the importer. The CCI and E shall issue the import license within 15 days of the receipt of the recommendations from DAC. However the recommendation of DAC is not required for import of trial material.

(iv) **Quantity of Import:** A quantity of 5 kg of seed of coarse cereals/pulses and oilseeds for sowing sought to be imported by eligible importers, will be given to ICAR or such farms, which are accredited by ICAR, for trial and evaluation. After trial/evaluation for one crop season, the ICAR will intimate the results, agro-climatic zone-wise, within three months of the season to DAC. After the receipt of the results of the ICAR trial/evaluation, an eligible importer may apply for the import of such seed to the DAC.

On receipt of the application for commercial import, the DAC will consider the trial/evaluation report on the performance/resistance to seed and soil borne diseases of the seeds proposed to be imported.

(v) The characteristics of the varieties/hybrids cleared by DAC for import shall be 'given publicity through the AIR, Doordarshan and the extension agencies so that farmers will get to know the potential of the varieties/ hybrids cleared for import, as 'compared to the indigenous varieties/ hybrids. The decision to import or not to import the varieties/hybrids cleared by DAC will be left to the importers.

(vi) **Seed for Gene Bank:** All importers shall make available a small but specified quantity of the seed being imported to the ICAR at cost price, for testing/accession to the Gene Bank with the NBPGR.

(vii) The bulk import of seeds of coarse cereals, oilseeds and pulses shall be cleared/rejected by the PPA after quarantine checks within a period not exceeding three weeks. The rejected consignment shall be destroyed in accordance with the prescribed procedure.

(viii) The imported consignment shall be subjected to detailed testing for a period of 30-35 days. During quarantine the imported consignment shall be kept in a customs bonded warehouse at the cost of the importer.

Q4. What are guidelines for import of vegetable and flower seeds?

Ans. The availability of high yielding vegetable seeds is essential for increasing vegetable production and productivity. The additional production will generate higher incomes to the farmers and ensure the increased availability of vegetables at reasonable prices to the consumers.

(i) Further, there is a significant demand for fresh vegetables and flowers in the international market. The production, on a competitive basis, of vegetables and flowers of international standards has to be developed/ accelerated to increase our export earnings.

(ii) It is, therefore, necessary to supply to the farmers the best available seeds of vegetables/flowers and ornamental plants. Accordingly, the import of seeds of vegetables/flowers and ornamental plants, tubers and bulbs, cuttings, saplings, bud woods, *etc.* of flowers will be allowed on OGL by identified categories.

(iii) The bulk import of vegetable and flower seeds is permitted under Open General license (OGL). The following organizations or importers shall be eligible to import these seeds/planting materials:

(1) Departments of Agriculture/Horticulture of the State Government, State Agriculture Universities and ICAR;

(2) Seed producing Indian companies/firms, after registration with the National Seeds Corporation;

(3) National Seeds Corporation, State Seeds Corporations;

(4) Food processing industrial units;

(5) Growers of vegetables and flowers registered with the Director of Agriculture/Horticulture of the State Government.

(iv) No post entry quarantine (PEQ) shall be necessary for the import of seeds of vegetables/flowers/ornamental plants and tubers and bulbs of flowers. The import will be permitted after visual inspection, fumigation, laboratory inspection and grow-out tests.

(v) The Plant Protection Adviser to the Government of India (PPA) may reject/clear the consignment of seeds of vegetables/flowers/ornamental plants and tubers and bulbs of flowers within a period not exceeding 50 days. The consignment shall be kept in a bonded warehouse at the cost of the importer until quarantine/customs clearance is given. The customs clearance of seeds of vegetables/flowers/ornamental plants and tubers and bulbs of flowers will be given by the customs authorities only after the quarantine clearance is given by the PPA.

The imported seed is subject to detailed seed testing for a period of 30-35 days on arrival at the port of entry. However, permit issued by PPA or any other competent officer notified for this purpose, is required for import of vegetable seeds but not flower seeds.

Q5. Describe briefly guidelines for import of bulbs and tubers of flowers.

Ans. (i) The bulk import of bulbs and tubers of flowers and ornamentals is allowed under OGL by eligible importers as stated above.

(ii) The imported consignments shall be subjected to grow-out test for a period of 35-40 days on arrival at the port of entry and for the same purpose the consignment will be held under detention in cold storage under customs bond or the imported bulbs and tubers may be subjected to post-entry quarantine instead of grow-out test at the specific request of the importer

(iii) The bulbs/tubers are required to be sown in individual poly bags and are subjected to joint inspection by designated Inspection Authority (DIA) and the officer of concerned plant quarantine (PQ) station during post entry quarantine (PEQ) period specified at the time of issue of permit.

Q6. Explain briefly guidelines for import of cuttings/saplings/bud wood *etc.*, of flowers.

Ans. (i) The import of cuttings/saplings/bud wood *etc.*, of flowers is permitted for import under Open General license (OGL).

(ii) A permit issued by PPA or any other competent officer notified for this purpose, is required. The importer is required to establish PEQ facilities prior to import, which are to be approved by the Designated Inspection Authorities (DIAs) as per the guidelines issued by PPA.

(iii) The imported consignments on arrival at the port of entry shall be subject to quarantine inspection and cleared within 24-72 hrs with a condition for growing under post-entry quarantine for a period not exceeding 45 days in an approved PEQ facility under the supervision of DIA.

For import of cuttings, saplings, bud wood *etc.* of flowers, PEQ shall be essential. PEQ facilities as may be prescribed by the PPA, shall have to be established by the eligible categories of importers permitted to import cuttings, saplings, bud wood *etc.* of flowers under OGL. The fact that such a facility has been established by him under the Plants, Fruits and Seeds (Regulation of import into India) Order, 1984. This agency, hereinafter referred to as the designated inspection agency (DIA), which may be the State Agriculture University (SAU), will also be authorized to carry out the PEQ inspection. The DIA's certificate that the importer has established the prescribed PEQ facility shall be presented by the eligible importer to the PPA before the imported consignment is given quarantine clearance by the PPA.

(i) The quarantine clearance subject to the PEQ condition will be given by the PPA after conducting visual inspection/laboratory inspection to see that the consignment of the imported cuttings, saplings, bud wood *etc.* of flowers, is free from any exotic pests or diseases.

(ii) Customs clearance shall be given by the customs authorities to the consignment on receipt of the quarantine clearance with PEQ condition from PPA. Approval of PEQ facility Guidelines to designated inspection agency for PEQ Samples to ICAR/Gene Bank Modification in REP Policy Import Policy for Seeds/Planting Introduction of pests and pathogens

(ii) Cuttings, saplings, bud woods *etc.* of flowers, which require PEQ inspection shall be grown in the PEQ facility approved by the PPA or a DIA authorized by the PPA. The period for which the cuttings, saplings, bud woods *etc.* shall be grown in such a facility shall be specified in the import permit to be granted by the PPA.

(iv) Detailed guidelines for the DIA (Appendix A) for conducting post entry quarantine shall be laid down by the PPA from time to time. If the PEQ reveal any exotic insect pest or disease the material shall be forthwith destroyed in the prescribed manner by the DIA under intimation to PPA.

(v) Small quantities of n flower/vegetable seeds tubers and cuttings and saplings will be made available by the importers to ICAR at cost price for testing/accession to the Gene Bank.

(vi) The import of seeds of vegetables, flowers, ornamental plants and bulbs and tubers of flowers, will continue to be allowed under REP. However, the transfer of REP licenses for items of seeds of vegetables/flowers/ornamental plants, bulbs and tubers of flowers will be allowed only to those persons, who are otherwise eligible to import these items under OGL subject to the conditions laid down in the Plants, Fruits and Seeds (Regulation of Import into India) Order, 1984.

(vii) The import of seeds/planting material for fruits shall be permitted selectively by DAC, on a case-to-case basis, on the recommendation of the State Director of Horticulture/Agriculture, subject to such quarantine regulations as may be laid down by the PPA.

Q7. Discuss briefly guidelines for import of seeds/planting material of fruits.

Ans. (i) The seeds/planting material of fruit plant species are selectively permitted for import by DAC on case-to-case basis on the recommendation of Director of Horticulture/Agriculture of the state and subject to quarantine regulations as may be laid down by PPA.

(ii) However, permit issued by PPA or any other competent officer notified for this purpose, is required for the import of the same.

Q8. **Discuss briefly guidelines for strengthening of plant quarantine as per New Seed Policy 1988.**

Ans. (1) In the past, several pests and pathogens have been introduced through seed material imported into the country. Many of these pests have since become wide-spread. Some examples are downy mildew on sunflower, spotted wilt virus on tomato, peanut stripe virus on groundnut, apple scab from Europe, golden nematode and wart diseases of potato from Europe and bacterial blight of paddy from East Asia, to mention only a few.

(2) Therefore, while importing seeds and planting materials, it has to be ensured that there is absolutely no compromise on plant quarantine procedures. Every effort has to be made to prevent the entry into India of exotic pests, diseases and weeds, which are detrimental to the interests of the farmers. The new policy on imports for seeds can thus be implemented fully only after the plant quarantine facilities are adequately strengthened within a specified time frame.

(3) There are 26 quarantine and fumigation stations located at 10 airports, 9 seaports and 7 land frontiers. It is proposed to allow import of seed/ planting material *etc.* initially only at Delhi, Bombay, Calcutta, Madras and Amritsar.

Therefore, the plant quarantine facilities at these five centres are being strengthened/modernized.

(4) The strengthening of the plant quarantine organization through the augmentation of technical staff, provision of physical infrastructure, essential laboratory equipment and the creation of regional offices for issue of import permits shall be a time-bound operation and shall be completed within a period of 12 months.

(5) The new policy on import of seed shall come into effect from the 1st October, 1988.

(6) The imports under the new policy shall be allowed, based on the procedures being followed by PPA and subject to the existing physical facilities of the Plant Quarantine and Fumigation Stations to undertake laboratory tests, fumigation *etc.* for a period of three months *i.e.* till the 31st December, 1988.

(7) During this 3 months period, PPA, while strengthening the quarantine facilities for grow-out tests available at the IARI, New Delhi and the SAUs or other research laboratories near Bombay, Calcutta, Madras and Amritsar. By the 1st of October, 1999, based on the time-bound program, the plant quarantine facilities at the 5 stations should have been augmented to implement the new policy in full. Pending this, from the 1st January to the 1st October, 1989, the implementation of the new policy will be phased/regulated, based on the progress in the commissioning of the new plant quarantine facilities at these five stations.

(8) A National Plant Quarantine Advisory Committee shall be constituted to advise the PPA on various technical matters in different disciplines of Entomology, Plant Pathology, Nematology, Virology, Bacteriology, Weed Sciences and for serological detection techniques of viruses. This Committee may meet at least twice a year.

Q9. Discuss briefly guidelines and procedure for quarantine/post entry quarantine as per New Seed Policy 1988.

(1) A separate Plant Quarantine Counter shall be provided in the arrival hall at Delhi, Bombay, Calcutta, Madras and Amritsar airports and at Bombay, Calcutta and Madras sea ports for effectively checking the imported seeds and planting materials and ensuring strict observance of plant quarantine procedure.

(2) All consignments of seeds and plants for sowing shall be imported into India in accordance with the provisions of Plants, Fruits and Seeds (Regulation of Import into India) Order, 1984. An import permit and a phytosanitary certificate with additional declaration, if necessary, as prescribed by the PPA, shall be essential. The pre-entry requirements for flowers and ornamental plants, vegetable seeds (FVS), bulbs and tubers (BT) and cuttings and saplings (CS) would be as follows:

   i) Import permit

   ii) Phytosanitary Certificate

   iii) Additional Declaration

      (a) Freedom from Soil

      (b) Freedom from Weeds

      (c) Plant Quarantine objects and

   iv) Approval of PEQ facilities

(3) After the arrival of consignments at the port of entry, quarantine checks would be undertaken which may include visual inspection, laboratory inspection, fumigation and grow out tests. For the purpose of these checks samples will be

Post Entry Quarantine

Sampling Procedures

Inspection requirements drawn and the tests will be conducted concurrently.

(4) PEQ will be required only for growing imported cuttings, saplings, budwood *etc.* of flowers. The importers will be required to create PEQ facilities, as may be prescribed by the PPA. The guidelines for PEQ facilities for importers are at Appendix 'B'. The planting materials, which require PEQ inspection, shall be grown in the PEQ facility approved by the PPA or the DIA in the State, authorized by the PPA under the provisions of the Plants, Fruits and Seeds (Regulation of Import into India) Order, 1984.

The period, for which the planting materials shall be grown in such a facility, shall be specified in the import permits granted by PPA. On the production of the certificate from the DIA that the importer has established the prescribed PEQ facility, the PPA will give the quarantine clearance after visual tests, laboratory tests, *etc.*

(5) The sampling procedures for various categories of imported vegetable/flower planting materials would be as under:

   a) FVS The sampling techniques shall be as per International Seed Testing Association (ISTA) rules and guidelines.

   b) BT and CS The minimum sample size is considered to be 10 numbers. However, in case of bulk imports, the representative sample for V.I. and L.I. would not be less than 0.1 per cent.

(6) The inspection under the categories mentioned above would comprise the following specific investigations/parameters:

(7) The technique for undertaking grow out tests has been prescribed by the PPA. The seeds of vegetables/flowers/ornamental plants after visual inspection shall be subject to fumigation at the cost of the importer to check pest infestation/disease before undertaking laboratory inspection.

(8) If the imported consignment shows the presence or manifestation of any exotic disease, pests or nematodes, the entire consignment shall be rejected and disposed of in accordance with the prescribed methods of destruction.

(9) The imported consignment showing presence of pests, diseases and nematodes recorded in India shall be released only after effective quarantine measures are undertaken to bring down the incidence to levels prescribed; otherwise, such consignments shall be disposed of as prescribed by the PPA. 6.10 If the samples drawn from the imported consignments satisfy the prescribed checks, the consignments shall be released for entry.

Q10. What are incentives to domestic seed industry as per New Seed Policy 1988?

Ans. (1) In the context of the liberalization of the import of seed/planting materials, incentives are proposed to be provided to domestic seed industry to encourage and promote its growth and enable it to produce viable high yielding varieties

(2) Accordingly, it has been decided to reduce the import duty on seeds to 15 per cent. The import duty reduction concession shall be accorded only to seeds meant for sowing and to planting materials.

(3) It is proposed to reduce the import duty on machines and equipments used for seed production/processing, quality control machinery *etc.*, not manufactured in the country or for which the technology upgrading is necessary.

(4) Pre-shipment credit up to 180 days will be allowed at 9.5 per cent rate of interest per annum. Beyond 180 days, and in all up to 270 days with the prior approval of the Reserve Bank of India, the rate of interest shall be 11.5 per cent per annum.

Post-shipment credit shall also be allowed at 9.5 per cent rate of interest per annum. A circular specifically allowing these facilities to seed exploring units will be issued by the Reserve Bank of India.

(5) Cash compensatory support on the export of seed will be considered by the Ministry of Commerce on the basis of data supplied by the seed exporting companies.

(6) Income tax rebate/deduction will be given to the seed industry on the revenue expenditure incurred on in-house research and development, to such units which pay income tax.

(7) Incentives, as provided to industries located in the backward areas/growth centres, may be, sought by DAC for seed industry. Such concessions for seed industry will be identified by DAC and taken up with the Ministry of Finance.

(8) A High Level Review Committee in the DAC will monitor the progress of the implementation of the Policy.

(9) The New Policy on Seed Development should help the seed industry to grow. The farmers will be benefited by getting access to the best available seed and planting material available anywhere in the world.

(10) The Government of India trusts that the State Governments./ Administrations of Union Territories/State Agricultural Universities will extend their fullest cooperation and assistance in the implementation of the Policy.

Q11. What are Guidelines for designated inspection agencies under the plants, fruits and seeds (regulation of import into India) order, 1984?

Ans. (1) Post Entry Quarantine (PEQ) shall be essential for imported cuttings, saplings, bud woods, *etc.* of flowers.

(2) Eligible categories of importers permitted to import cuttings, saplings, bud woods, *etc.* of flowers must establish post entry quarantine facilities, as may be prescribed by the PPA.

(3) A certificate shall be issued by the PPA or the Designated Inspection Agencies in the States authorized by the PPA under the Plants, Fruits and Seeds (Regulation of Import into India) Order, 1984 for conducting PEQ inspections, that the importer has established the prescribed PEQ facilities, after verification and evaluation of the PEQ facilities created by the importer.

(4) Imported cuttings, saplings, bud woods *etc.* of flowers shall be grown in the prescribed PEQ facility for a period specified in the import permit granted by the PPA.

(5) The DIA shall be located in the State Agricultural University.

(6) A multidisciplinary team of the DIA consisting of a Plant Pathologist/ Virologist/Entomologist shall inspect the imported planting material at the time of planting in the specified PEQ facility.

(7) Post planting observations shall be taken in the following manner: Cuttings: The first observation would depend upon the growth of the plant but normally it should be undertaken within 20-25 days followed by second observation within 40-45 days of planting.

Saplings: The first observation after planting shall depend upon the growth of the plant, which may normally be within 2-3 weeks.

During inspection, special notice should be taken of symptoms of any plant diseases, especially systemic and viral disease.

Disease affected plants should be destroyed in the presence of the DIA, as prescribed by PPA, to avoid any possible contamination by pathogens of the surrounding environment and plants.

In the event of observance of any exotic pest or disease, the material shall be forthwith destroyed in the prescribed manner under intimation to PPA.

(8) The final report of PEQ inspection shall be sent to PPA for information and record within 15 days of the final observation.

Q12. Explain Guidelines for importers for establishing post entry quarantine facilities.

Ans. (1) The importers of cutting and saplings, *etc.* of flowers will be required to establish post entry quarantine facilities, as prescribed by the Plant Protection Adviser. These facilities may include the establishment of a glass house/plastic house/any other specified in a defined area, in isolation.

(2) Imported cuttings, saplings, bud woods, *etc.* of flowers shall be grown in the Prescribed PEQ facility for a period specified in the import permit granted by the PPA.

(3) The importer shall facilitate the Designated Inspection Agency (DIA) to undertake inspection at specified intervals of time.

(4) The importer shall inform the DIA about the planting dates well in advance. If there is any occurrence of pest or disease symptoms, he shall immediately inform the DIA. The importer shall abide by the decision of the DIA.

5. No plant or parts thereof whether living or dead shall be removed during the pendency of post entry quarantine without the approval of the DIA.

(6) A qualified Plant Pathologist/Entomologist/Plant Virologist/ Horticulturist should be engaged as consultant at the cost of the importer for looking after the plants under post entry quarantine.

(7) On the detection of any exotic insect pest or disease, the entire imported material shall be destroyed forthwith in a prescribed manner under intimation to PPA.

(8) During inspection, the multidisciplinary team of the DIA shall take special notice of symptoms of any plant diseases specially systemic and viral diseases. The disease affected plants shall be destroyed in the presence of DIA to avoid possible contamination of the surrounding plants and environment by pathogens and the soil shall be disinfected with formalene.

(9) The importer shall pay the cost of post entry quarantine inspection, as prescribed by PPA. He shall make the deposit, as may be laid down, on the quarantine clearance of the consignment by the PPA. The consignment shall be released only after the importer makes the payment in full of the DIA inspection fees to the PPA.

(10) The importer shall maintain basic inspection tools like hand lenses, microscopes, chemicals, dissection box, glass vials and petri plates.

(11) A log book shall be maintained for proper record for incoming and out going material as well as the dates of inspection and detailed report by the DIA.

(12) A register containing details of imported material, panting date, manifestation of diseases and appearance of pests and control measures taken shall be maintained.

(13) The PEQ areas shall have the provision of locking arrangements for security purposes.

Chapter 7

Seed Order, 1989

Q1. What is Seed Order, 1989?

Ans. The seed order, 1989 was approved by the government of India, Ministry of Agriculture (Department of Agriculture and Cooperation) on 27th October, 1989. This order suppresses the Plants, Fruits and Seeds (Regulation of import into India) Order 1984. This order may be called the Plants, Fruits and Seeds (Regulation of Import into India) Order, 1989. This order consists of six chapters. Chapter 1 deals with title, scope and definitions of various terms. Chapter 2 deals with general conditions of import. Chapter 3 explains special conditions of import. Chapter 4 deals with post entry quarantine, Chapter 5 deals with appeals and revision and chapter 6 explains powers of relaxation.

Q2. Explain general conditions for import as per seed order, 1989.

Ans. All consignments of plants, fruits and seeds shall be imported into India subject to the following conditions.

(1) **Valid Permit:** A valid permit is compulsory for import of any consignment into India.

(2) **Application:** An application for a permit to import consignments by land, air or sea has to be sent in triplicate) at least one month in advance to the Competent Authority. The application for the import of seeds, fruits and Plants for consumption shall be made in form 'A' and that for the import of seeds and plants for sowing or planting shall be made in form 'B'.

(3) **Fees:** A fee of Rs.50 is to be paid along with the application for the import of seeds, fruits and plants for consumption and Rs.100 for application for the import of seeds and plants for sowing or planting. The fee is to be paid in the form of Demand Draft payable to the Competent Authority having jurisdiction.

(4) Issue of Permit: If the application is in order, the Competent Authority shall issue permit in Form "C" for import of seeds and fruits for consumption and in Form "D" for import of seeds and plants for sowing or planting. If application does not meet requirements, the issue of permit may be refused or withheld by the Competent Authority.

(5) Validity of Permit: The import permit issued for such import shall be valid for a period of six months. However, the Competent Authority may, on request, extent the period of validity for a further period of six months, for genuine reasons.;

(6) Tag Color: The Competent Authority shall forward to the importer an orange and green color tag specified in form "E", in the case of permits issued for import of seeds and plants for sowing or planting so as to facilitate the identification of consignments at the time of their arrival at the land customs station or port of entry.

(7) Entry Points: All the consignments for consumption, sowing and propagation or planting shall be imported into India only through entry points notified by the Central Government from time to time in this behalf. However, all consignments of dry fruits, fresh fruits and vegetables for consumption, imported from Afghanistan, Pakistan and a West Asian Countries by land shall be imported only through Attari-Wagha Border.

(8) All consignments of plants and seeds for sowing and propagation or planting shall be imported into India through land customs stations, seaport, airport at Amritsar, Bombay, Calcutta, Delhi and Madras and such other entry points as may be specifically notified by the Central Government from time to time.

(9) Inspection of Material: The consignment, on arrival, at an entry point, shall be inspected by the Plant Protection Adviser or any other officer duly authorized by him in this behalf, in accordance with the guidelines issued by the Plant Protection Adviser from time to time.

(10) The Plant Protection Adviser or the officer authorized by him may, after inspection, fumigation, disinfection or imported for personal consumption may be allowed to be imported without a Phytosanitary to the country of origin;

(11) Cost of Fumigation, *etc.*: Where fumigation or disinfestation or the; disinfection is considered necessary in respect of a consignment of plants, seeds and fruits of more than 1000 cubic metre in volume, the importer shall on his own or at his cost through an agency approved by the Plant Protection Adviser arrange for the fumigation, disinfection or disinfection of the consignment, under the supervision of an officer duly authorised by the Plant Protection Adviser in that behalf:

(12) Responsibility of Importer: The importer has the following responsibility.

 (a) To bring the consignments to the concerned Plant Quarantine and Fumigation Station, or to places of inspection, fumigation or treatment as directed by the Plant Protection Adviser or the officer duly authorized by him;

(b) To open, repack and load into or unload from the fumigation chamber and seal the consignments; and seal the consignments; and

(c) To remove them after inspection and treatment, according to the directions issued by the Plant Protection Adviser or an officer duly authorized by him.

The consignments intended for other countries shall be allowed transit through or transshipment at air or sea ports or land customs stations, provided they are packed in such a manner as will not permit spillage of any soil or materials or escape of any pest, and subject also to the condition that they are not opened in any place in India.

(13) Phytosanitary Certificate: No consignment shall be imported unless accompanied by an official Phytosanitary Certificate issued by the authorized officer of the country of origin of the consignment. Provided that cut flowers, garlands, bouquets, fruits and vegetables weighing less than two kilograms.

(14) Packaging: Consignments for import should be packed in the packaging materials envisaged as in clause 2(g) of this order. No consignment wherein hay or straw or any material of plant origin is used for packaging or as a part of packaging material shall be allowed to be imported.

(15) Import of Soil: Import of soil, earth, compost, sand, plant debris along with plants, fruits or seeds shall not be permitted except under the following conditions:

(a) The consignments of soil, earth, clay and similar material for any microbiological, soil-mechanics or minerological investigation and peat for horticultural may be permitted through specified air or sea ports or land custom station, on applications made for that purpose:

(b) The application for the purpose referred to in (I) above shall be made to the Plant Protection Adviser, at least one month in advance, in form "F"

(c) The Plant Protection Adviser may, after scrutiny of the application, and if satisfied of the purpose for which such consignment is being imported, issue special permit in Form "G"

(d) The consignments shall be inspected, fumigated disinfected or disinfested, on arrival, by the Plant Protection Adviser or any other officer duly authorized by him in this behalf.

(16) Fees: The importer of the consignments or his agent shall pay to the Plant Protection Adviser or any other officer duly authorized by him in this behalf, the fees prescribed in Schedule III to meet the cost of inspection, fumigation and disinfection before the release of the consignments.

Q3.   **What are special conditions for import as per seed order, 1989?**

Ans.   In addition to the general conditions of import of Plants, Fruits and Seeds, there are some special conditions for import of these commodities. The following items can be imported [Chapter 3].

    (i)   Coconut, seeds and all species of *Cocos;*

    (ii)   Coffee plants and seeds, and all species of Coffee;

    (iii)   Cotton seeds, and all species of *Gossypium;*

    (iv)   Seeds of forest trees;

    (v)   Groundnut seeds, and all species of *Araches.*

    (vi)   Bucrene and all species of *Medicago;*

    (vii)   Potato and all species of *Solanum;*

    (viii)   Rubber and all species of *Hevea;*

    (ix)   Sugarcane and all species of *Saccharum;*

    (x)   Tobacco and all species of *Nicotiana;*

    (xi)   Berseem and all species of *Trifolium;*

    (xii)   Sunflower and all species of *Helianthus;*

    (xiii)   Wheat and all species of *Triticum;*

    (xiv)   Paddy and all species of *Oryzae;*

    (xv)   Cuttings, saplings and bud-woods of flowers or ornamental plants;

    (xvi)   Seeds and the plant material of fruits.

Every consignment of the articles herein before mentioned shall be accompanied with the official Phytosanitary Certificate issued by the authorized officer of the country of origin of consignment, containing additional declarations that they are free from specified pests.

Q4.   **What are post-entry quarantine guidelines for import as per seed order, 1989?**

Ans.   Plants and seeds, which require Post-entry Quarantine, shall be grown in Post-entry Quarantine facilities, approved and certified by the Designated Inspection Authority, to conform to the conditions laid down by the Plant Protection Adviser. The period for which and the conditions under which, the plants and seeds shall be grown in such facilities shall be specified in the permit. The Post-entry Quarantine facilities shall be established and provided by the importer or his agent at his own cost and these shall be ready for use at the time of arrival of the consignment in India. The importer shall obtain a certificate from the Designated Inspection Authority who, after inspection of the Post-entry Quarantine facilities, shall certify that such Post-entry Quarantine facilities have been duly established and provided in accordance with the guidelines of the Plant Protection Adviser. The importer shall produce this certificate before the Officer-in-charge of the Quarantine Station at the entry point, at the time of arrival of the consignment

(i) The Officer-in-charge of the Quarantine Station. If after inspection of the consignment is satisfied, shall accord quarantine clearance with Post-entry-quarantine condition on the production, by an importer, of a certificate from the Designated Inspection Authority.

(ii) After according quarantine clearance with Post-entry Quarantine conditions to the consignments of plants and seeds requiring Post-entry Quarantine, the officer-in-charge of the Quarantine Station at the entry point shall inform the Designated Inspection Authority, having jurisdiction over the Post-entry Quarantine facility of their arrival at the location where such plants would be grown by the importer.

The importer shall inform in advance the Designated Inspection Authority having jurisdiction, about the time of planting of such materials.

The importer shall permit to the Designated Inspection Authority complete access to the Post-entry Quarantine facility for the inspection of plants and shall, at all times, abide by his instructions concerning the plants in the Post-entry Quarantine.

The Designated Inspection Authority shall inspect the plants grown in the Post-entry Quarantine facility of the importer for the detection of the pests and diseases.

(i) The Designated Inspection Authority shall permit the release of plants from Post-entry Quarantine, if they are found to be free from pests and diseases for the period specified in the permit for importation.

(ii) Where the plants in the Post-entry Quarantine are found to be affected by pests and diseases during the specified period:

    a. The Designated Inspection Authority shall order the destruction or return to the country of origin of the affected consignment of whole or a part of the plant population in the Post-entry Quarantine if the pest or disease is exotic and

    b. The Designated Inspection Authority shall advise the importer about the curative measures to be taken to the extent necessary, if the pest or disease is not exotic and permit the release of the affected population from the Post-entry Quarantine only after quarantine measures have been observed. Otherwise, the plants shall be ordered to be destroyed.

(iii) Where destruction of any plant population is ordered by the Designated Inspection Authority, the importer shall destroy the same in the prescribed manner under the supervision of Designated Inspection Authority.

(iv) 12. The importer shall be liable to pay the prescribed fee for inspection of plants in the Post-entry Quarantine facility as laid down in Schedule III.

Q5. Describe guidelines for appeal, revision and power of relaxation as per seed order, 1989?

Ans. If an importer is aggrieved by the decision of a Designated Inspection Authority regarding the destruction of any plant population, he may appeal to the Plant Protection Adviser within 7 days from the date of communication of the decision giving the grounds of appeal. It shall lawful for the Plant Protection Adviser to relay on the observation of the incidence of pests and diseases and Designated Inspection Authority and such other expert opinion, as he may deem necessary, for deciding the appeal.

The memorandum of appeal shall set out the grounds on which the decision is challenged and shall accompany with a fee of Rs. 10.00 by a Treasury Challan. The Plant Protection Adviser may, at any time, call for the records relating to any case pending before the Designated Inspecting Authority for the purpose of satisfying itself as to the legality or propriety of any decision passed by that authority and may pass such order in relating thereto, as it thinks fit:

No such order shall be passed after the expiry of three months from the date of the decision: Moreover, the Plant Protection Adviser shall not pass any order prejudicial to any person, without giving him a reasonable opportunity of hearing.

Powers of Relaxation: The central government may, in public interest, relax any of the conditions of this order relating to the permit and the phyto-sanitary certificate in relating to the import of any consignment.

Chapter 8

National Seeds Policy, 2002

Q1. What are aims and objectives of National Seeds Policy 2002?

Ans. The Seeds Act, 1966, Seeds Control Order, 1984 and the New Policy on Seeds Development, 1988, form the basis of promotion and regulation of the Seed Industry. In India, enactment of these seed legislations have made significant changes in economic and agricultural scenario. The main objectives of the National Seeds Policy are briefly presented as follows:

(i) **To Enhance Seed Replacement Rate:** It is necessary to enhance the seed replacement rates of various crops for achieving future targets of the food production.

(ii) **To Provide Quality Seed at Cheaper rate:** There is need to increase in the production of quality seeds, in which the private sector is expected to play a major role. Both private and Public Sector Seed Organizations at both Central and State levels, have to adopt economic pricing policies which would seek to realize the true cost of production.

(iii) **Rapid Development of Seed Industry:** The National Seed policy will create favorable climate for growth of a competitive and localized seed industry, The National Seeds Policy provides an appropriate climate for the seed industry to utilize available and prospective opportunities

(iv) **To encourage import of useful germplasm: The** National Seed policy will encourage import of useful germplasm that can be used for developing high yielding varieties and hybrids in different field crops.

(v) **Conservation of Agro-biodiversity: The** National Seed policy will safeguard the interests of Indian farmers and the conservation of agro-biodiversity.

(vi) **Safety from GMO to human and animal health: The** National Seed policy emphasizes on the safe use of genetically modified organisms avoiding possible harm to human and animal health.

(vii) **Use of Biotechnology for varietal improvement: The** National Seed policy provides conducive atmosphere for application of frontier sciences in varietal development and for enhanced investments in research and development. Globalization and economic liberalization have opened up new opportunities as well as challenges.

Q2. What are thrust areas of National Seeds Policy 2002?

Ans. There are several thrust areas of National Seeds Policy, 2002. Important thrust areas include (1) varietal development, (2) plant variety protection, (3) seed production, (4) quality assurance, (5) seed distribution and marketing, (6) infrastructure facilities, (7) transgenic plant varieties, (8) import of seeds and planting material, (9) export of seed, (10) promotion of domestic seed industry and (11) strengthening of monitoring system.

Q3. Describe varietal development in relation to National Seeds Policy 2002.

Ans. The development of new and improved varieties of plants and availability of such varieties to Indian farmers is of crucial importance for a sustained increase in agricultural productivity. Appropriate policy framework and programmatic interventions will be adopted to stimulate varietal development in tune with market trends, scientific-technological advances, suitability for biotic and abiotic stresses, location specific adaptability and farmers' needs.

Q4. Describe plant variety protection in relation to National Seeds Policy 2002.

Ans. An effective *sui generis* system for intellectual property protection will be implemented to stimulate investment in research and development of new plant varieties and to facilitate the growth of the Seed Industry in the country.

A Plant Varieties and Farmers' Rights Protection (PVP) Authority will be established which will undertake registration of extant and new plant varieties through the Plant Varieties Registry on the basis of varietal characteristics.

(i) **Criteria for Registration:** The registration of new plant varieties by the PVP Authority will be based on the criteria of novelty, distinctiveness, uniformity and stability.

(ii) The criteria of distinctiveness, uniformity and stability could be relaxed for registration of extant varieties, which will be done within a specified period to be decided by the PVP Authority.

(iii) **Registration:** Registration of all plant genera or species as notified by the Authority will be done in a phased manner.

(iv) Characterization: The PVP Authority will develop characterization and documentation of plant varieties registered under the PVP Act and cataloguing facilities for all varieties of plants.

(v) Farmers' Rights: The rights of farmers to save, use, exchange, share or sell farm produce of all varieties will be protected, with the proviso that farmers shall not be entitled to sell branded seed of a protected variety under the brand name.

(vi) Researchers' Rights: The rights of researchers to use the seed/planting material of protected varieties for bona-fide research and breeding of new plant varieties will be ensured.

(vii) Benefit Sharing: Equitable sharing of benefit arising out of the use of plant genetic resources that may accrue to a breeder from commercialization of seeds/planting materials of a new variety, will be provided.

(viii) Rewards: Farmers/groups of farmers/village communities will be rewarded suitably for their significant contribution in evolution of a plant variety subject to registration. The contribution of traditional knowledge in agriculture needs to be highlighted through suitable mechanisms and incentives.

(ix) Gene Fund: A National Gene Fund will be established for implementation of the benefit sharing arrangement, and payment of compensation to village communities for their contribution to the development and conservation of plant genetic resources and also to promote conservation and sustainable use of genetic resources. Suitable systems will be worked out to identify the contributions from traditional knowledge and heritage.

(x) Access to Germplasm: Plant Genetic Resources for Food and Agriculture Crops will be permitted to be accessed by Research Organizations and Seed Companies from public collections as per the provisions of the 'Material Transfer Agreement' of the International Treaty on Plant Genetic Resources and the Biological Diversity Bill.

(xi) Interaction between Public and Private Sector: Regular interaction amongst the Private and Public Researchers, Seed Companies/ Organizations and Development Agencies will be fostered to develop and promote growth of a healthy seed industry in the country.

(xii) Institutional Linkages: To keep abreast of global developments in the field of Plant Variety Protection and for technical collaboration, India may consider joining Regional and International Organizations.

(xiii) Compulsory Licensing: The PVP Authority may, if required, resort to compulsory licensing of a protected variety in public interest on the ground that requirements of the farming community for seeds and propagating material of a variety are not being met or that the production of the seeds or planting material of the protected variety is not being facilitated to the fullest possible extent.

Q5. Describe seed production in relation to National Seeds Policy 2002.

Ans. To meet the Nation's food security needs, it is important to make available to Indian farmers a wide range of seeds of superior quality, in adequate quantity on a timely basis. Public Sector Seed Institutions will be encouraged to enhance production of seed towards meeting the objective of food and nutritional security.

   (i) **Categories of Seed:** The Indian seed program adheres to the limited three generation system of seed multiplication, namely, breeder, foundation and certified seed.

   (ii) Nucleus Seed: It is the seed produced by the breeder to develop the particular variety and is directly used for multiplication as breeder seed.

   (iii) **Breeder Seed:** Breeder seed is the progeny of nucleus seed. Breeder seed is the seed material directly controlled by the originating or the sponsoring breeder or Institution for the initial and recurring production of foundation seed.

   (iv) **Foundation Seed:** It is the progeny of breeder seed. Foundation seed may also be produced from foundation seed. Production of foundation seed stage-I and stage-II may thus be permitted, if supervised and approved by the Certification Agency and if the production process is so handled as to maintain specific genetic purity and identity.

   (v) **Certified seed:** It is the progeny of foundation seed or the progeny of certified seed. If the certified seed is the progeny of certified seed, then this reproduction will not exceed three generations beyond foundation stage-I and it will be ascertained by the Certification Agency that genetic identity and genetic purity has not been significantly altered.

   (vi) **Access to Breeder Seed:** Public Sector Seed Production Agencies will continue to have free access to breeder seed under the National Agriculture Research System. The State Farms Corporation of India and National Seeds Corporation will be restructured to make productive use of these organizations in the planned growth of the Seed Sector. Private Seed Production Agencies will also have access to breeder seed subject to terms and conditions to be decided by Government of India.

   (vii) **Responsibility of Seed Production:** State Agriculture Universities/ICAR Institutes will have the primary responsibility for production of breeder seed as per the requirements of the respective States.

   (viii) **Seed Quality:** Special attention will be given to the need to upgrade the quality of farmers' saved seeds through interventions such as the Seed Village Scheme.

   (ix) **Rate of Seed Replacement:** Seed replacement rates will be raised progressively with the objective of expanding the use of quality seeds.

(x) **National Seed Map:** DAC, in consultation with ICAR and States, will prepare a National Seed Map to identify potential, alternative and non-traditional areas for seed production of specific crops.

(xi) **Perspective Plan:** To put in place an effective seed production program, each State will undertake advance planning and prepare a perspective plan for seed production and distribution over a rolling (five to six year) period. Seed Banks will be set up in non-traditional areas to meet demands for seeds during natural calamities.

(xii) **Seed Village Scheme:** The 'Seed Village Scheme' will be promoted to facilitate production and timely availability of seed of desired crops/varieties at the local level. Special emphasis will be given to seed multiplication for building adequate stocks of certified/quality seeds by providing foundation seed to farmers.

(xiii) **Popularizing New Varieties:** For popularizing newly developed varieties and promoting seed production of these varieties, seed mini-kits of pioneering seed varieties will be supplied to farmers. Seed exchange among farmers and seed producers will be encouraged to popularize new/non-traditional varieties. Seeds of newly developed varieties must be made available to farmers with minimum time gap. Seed producing agencies will be encouraged to tie up with Research Institutions for popularization and commercialization of these varieties.

(xiv) **Support for Hybrid Seed Production:** As hybrids have the potential to improve plant vigor and increase yield, support for production of hybrid seed will be provided.

(xv) **Seed Production in New Areas:** Seed production will be extended to agro-climatic zones which are outside the traditional seed growing areas, in order to avoid non-remunerative seed farming in unsuitable areas.

(xvi) **Establishment of Seed Banks:** Seed Banks will be established for stocking specified quantities of seed of required crops/varieties for ensuring timely and adequate supply of seeds to farmers during adverse situations such as natural calamities, shortfalls in production, *etc.* Seed Banks will be suitably strengthened with cold storage and pest control facilities.

(xvii) **Seed Storage:** The storage of seed at the village level will be encouraged to facilitate immediate availability of seeds in the event of natural calamities and unforeseen situations. For the storage of seeds at farm level, scientific storage structures will be popularized and techniques of scientific storage of seeds will be promoted among farmers as an extension practice.

(xviii) **Seed Crop Insurance:** Seed growers will be encouraged to avail of Seed Crop Insurance to cover risk factors involved in production of seeds. The Seed Crop Insurance Scheme will be reviewed so as to provide effective risk cover to seed producers and will be extended to all traditional and non-traditional areas covered under the seed production programme.

Q6. Describe seed quality assurance in relation to National Seeds Policy 2002.

Ans. The Seeds Act will be revised to regulate the sale, import and export of seeds and planting materials of agriculture crops including fodder, green manure and horticulture and supply of quality seeds and planting materials to farmers throughout the country.

    (i) Establishment of National Seeds Board: The National Seeds Board (NSB) will be established in place of existing Central Seed Committee and Central Seed Certification Board. The NSB will have permanent existence with the responsibility of executing and implementing the provisions of the Seeds Act and advising the Government on all matters relating to seed planning and development. The NSB will function as the apex body in the seed sector.

    (ii) Registration: All varieties, both domestic and imported varieties, that are placed on the market for sale and distribution of seeds and planting materials will be registered under the Seeds Act. However, for vegetable and ornamental crops a simple system of varietal registration based on "breeders declaration" will be adopted.

    (iii) The Board will undertake registration of kinds/varieties of seeds that are to be offered for sale in the market, on the basis of identified parameters for establishing value for cultivation and usage (VCU) through testing/trialing.

    (iv) Period of Registration: Registration of varieties will be granted for a fixed period on the basis of multi-location trials to determine VCU over a minimum period of three seasons, or as otherwise prescribed as in the case of long duration crops and horticultural crops. Samples of the material for registration will be sent to the NBPGR for retention in the National Gene Bank.

    (v) Varieties that are in the market at the time of coming into force of the revised Seeds Act, will have to be registered within a fixed time period, and subjected to such testing as will be notified.

    (vi) The NSB will accredit ICAR, SAUs, public/private organizations to conduct VCU trials of all varieties for the purpose of registration as per prescribed standards.

    (vii) The NSB will maintain the National Seeds Register containing details of varieties that are registered. This will help the Board to coordinate and assist activities of the States in their efforts to provide quality seeds to farmers.

    (viii) The NSB will prescribe minimum standards (of germination, genetic characteristics, physical purity, seed health, *etc.*) as well as suitable guidelines for registration of seed and planting materials.

(ix) Provisional registration would be granted on the basis of information filed by the applicant relating to trials over one season to tide over the stipulation of testing over three seasons before the grant of registration.

(x) Government will have the right to exclude certain kinds or varieties from registration to protect public order or human, animal and plant life and health, or to avoid serious prejudice to the environment.

(xi) The NSB will have the power to cancel the registration granted to a variety if the registration has been obtained by misrepresentation or concealment of essential data, the variety is obsolete and has outlived its utility and if the prevention of commercial exploitation of such variety is necessary in the public interest.

(xii) Registration of Seed Processing Units will be required if such Units meet the prescribed minimum standards for processing the seed.

(xiii) Seed Certification will continue to be voluntary. The Certification tag/ label will provide an assurance of quality to the farmer.

(xiv) The Board will accredit individuals or organizations to carry out seed certification including self-certification on fulfillment of criteria as prescribed.

(xv) To meet quality assurance requirements for export of seeds, Seed Testing facilities will be established in conformity with ISTA and OECD seed certification programs.

(xvi) The State Government, in conformity with guidelines and standards specified by the Board, will establish one or more State Seed Testing Laboratories or declare any Seed Testing Laboratory in the Government or non-Government Sector as a State Seed Testing Laboratory where analysis of seeds will be carried out in the prescribed manner.

(xvii) Farmers will be encouraged to use certified seeds to ensure improved performance and output.

(xviii) **Farmers' Rights:** Farmers will retain their right to save, use, exchange, share or sell their farm seeds and planting materials without any restriction. They will be free to sell their seed on their own premises or in the local market without any hindrance provided that the seed is not branded. Farmers' right to continue using the varieties of their choice will not be infringed by the system of compulsory registration.

(xix) **Strict Measures:** Stringent measures would be taken to ensure the availability of high quality of seeds and check the sale of spurious or misbranded seeds.

Q7. Describe seed distribution and marketing in relation to National Seeds Policy 2002.

Ans. The availability of high quality seeds to farmers through an improved distribution system and efficient marketing set-up will be ensured to facilitate greater security of seed supply.

(i) For promoting efficient and timely distribution and marketing of seed throughout the country, a supportive environment will be provided to encourage expansion of the role of the private seed sector. Efforts will be made to achieve better coordination between State Governments to facilitate free Inter-State movement of seed and planting material through exemption of duties and taxes.

(ii) Private Seed Sector will be encouraged and motivated to restructure and reorient their activities to cater to non-traditional areas.

(iii) A mechanism will be established for collection and dissemination of market intelligence regarding preference of consumers and farmers.

(iv) A National Seed Grid will be established as a data-base for monitoring of information on requirement of seed, its production, distribution and preference of farmers on a district-wise basis.

(v) Access to term finance from Commercial Banks will be facilitated for developing efficient seed distribution and marketing facilities for growth of the seed sector.

(vi) Distribution and marketing of seed of any variety, for the purpose of sowing and planting will be allowed only if the said variety has been registered by the National Seeds Board.

(vii) National Seeds Board can direct a dealer to sell or distribute seeds in a specified manner in a specified area if it is considered necessary to the public interest.

Q8. Describe development of infrastructure facilities in relation to National Seeds Policy 2002.

Ans. To meet the enhanced requirement of quality/certified seeds, creation of new infrastructure facilities along with strengthening of existing facilities, will be promoted.

National Seed Research and Training Center will be set up to impart training and build a knowledge base in various disciplines of the seed sector.

(i) The Central Seed Testing Laboratory will be established at the National Seed Research and Training Center to perform referral and other functions as required under the Seeds Act.

(ii) Seed processing capacity will be augmented to meet the enhanced requirement of quality seed.

(iii) Modernization of seed processing facilities will be encouraged in terms of modern equipment and latest techniques, such as seed treatment for enhancement of performance of seed, *etc.*

(iv) Conditioned storage for breeder and foundation seed and aerated storage for certified seed would be created in different regions.

(v) A computerized National Seeds Grid will be established to provide information on availability of different varieties of seeds with production agencies, their location, quality *etc.* This network will facilitate optimum utilisation of available seeds in every region.

(vi) Initially, seed production agencies in the public sector would be connected with the National Seed Grid, but progressively the private sector will be encouraged to join the Grid for providing a clear assessment of demand and supply of seeds.

(vii) State Governments, or the National Seeds Board in consultation with the concerned State Government, may establish Seed Certification Agencies.

(viii) State Governments will establish appropriate systems for effective execution and implementation of the objectives and provisions of the Seeds Act.

Q9. Describe transgenic plant varieties in relation to National Seeds Policy 2002.

Ans. Biotechnology will play a vital role in the development of the agriculture sector. This technology can be used not only to develop new crops/varieties, which are tolerant to disease, pests and abiotic stresses, but also to improve productivity and nutritional quality of food.

(i) All genetically engineered crops/varieties will be tested for environment and biosafety before their commercial release, as per the regulations and guidelines of the Environment Protection Act (EPA), 1986.

(ii) The EPA, 1986, read with the Rules, 1989 would adequately address the safety aspects of transgenic seeds/planting materials. A list will be generated from Indian experience of transgenic cultivars that could be rated as environmentally safe.

(iii) Seeds of transgenic plant varieties for research purposes will be imported only through the National Bureau of Plant Genetic Resources (NBPGR) as per the EPA, 1986.

(iv) Transgenic crops/varieties will be tested to determine their agronomic value for at least two seasons under the All India Coordinated Project Trials of ICAR, in coordination with the tests for environment and biosafety clearance as per the EPA before any variety is commercially released in the market.

(v) After the transgenic plant variety is commercially released, its seed will be registered and marketed in the country as per the provisions of the Seeds Act.

(vi) After commercial release of a transgenic plant variety, its performance in the field,will be monitored for at least 3 to 5 years by the Ministry of Agriculture and State Departments of Agriculture.

(vii) Transgenic varieties can be protected under the PVP legislation in the same manner as non-transgenic varieties after their release for commercial cultivation.

  (viii) All seeds imported into the country will be required to be accompanied by a certificate from the Competent Authority of the exporting country regarding their transgenic character or otherwise.

  (ix) If the seed or planting material is a product of transgenic manipulation, it will be allowed to be imported only with the approval of the Genetic Engineering Approval Committee (GEAC), set up under the EPA, 1986.

  (x) Packages containing transgenic seeds/planting materials, if and when placed on sale, will carry a label indicating their transgenic nature. The specific characteristics including the agronomic/yield benefits, names of the transgenes and any relevant information shall also be indicated on the label.

  (xi) Emphasis will be placed on the development of infrastructure for the testing, identification and evaluation of transgenic planting materials in the country.

Q10. Describe import of seeds and planting material in relation to National Seeds Policy 2002.

Ans. The objective of the import policy is to provide the best planting material available anywhere in the world to Indian farmers, to increase productivity, farm income and export earnings, while ensuring that there is no deleterious effect on environment, health and biosafety.

  (i) While importing seeds and planting material, care will be taken to ensure that there is absolutely no compromise on the requirements under prevailing plant quarantine procedures, so as to prevent entry into the country of exotic pests, diseases and weeds detrimental to Indian agriculture.

  (ii) All imports of seeds will require a permit granted by the Plant Protection Advisor to the Government of India, which will be issued within the minimum possible time frame.

  (iii) All import of seeds and planting materials, *etc.* will be allowed freely subject to EXIM Policy guidelines and the requirements of the Plants, Fruits and Seeds (Regulation of import into India) Order, 1989 as amended from time to time. Import of parental lines of newly developed varieties will also be encouraged.

  (iv) Seeds and planting materials imported for sale into the country will have to meet minimum seed standards of seed health, germination, genetic and physical purity as prescribed.

  (v) All importers will make available a small sample of the imported seed to the Gene Bank maintained by NBPGR.

  (vi) The existing policy, which permits free import of seeds of vegetables, flowers and ornamental plants, cuttings, saplings of flowers, tubers and bulbs of flowers by certain specified categories of importers will continue. Tubers and bulbs of flowers will be subjected to post-entry quarantine.

(vii) After the arrival of consignments at the port of entry, quarantine checks would be undertaken; which may include visual inspection, laboratory inspection, fumigation and grow-out tests. For the purpose of these checks, samples will be drawn and the tests will be conducted concurrently.

Q11. Describe export of seeds in relation to National Seeds Policy 2002.

Ans. Due to diversity of agro-climatic conditions, strong seed production infrastructure and market opportunities, India holds significant promise for export of seeds.

  (i) Government will evolve a long term policy for export of seeds with a view to raise India's share of global seed export from the present level of less than 1 per cent to 10 per cent by the year 2020.

 (ii) The export policy will specifically encourage custom production of seeds for export and will be based on long term perspective, dispensing with case to case consideration of proposals.

(iii) Establishment and strengthening of Seeds Export Promotion Zones with special incentives from the Government will be facilitated.

(iv) A data bank will be created to provide information on the International Market and on export potential of Indian varieties in different parts of the world.

 (v) A database on availability of seeds of different crops to assess impact of exports on domestic availability of seeds will be created.

(vi) Promotional programmes to improve the quality of Indian seeds to enhance its acceptability in the International Market will be taken up.

(vii) Testing and certification facilities will be established in conformity with international requirements.

Q12. Describe promotion of domestic seed industry in relation to National Seeds Policy 2002.

Ans. Incentives will be provided to the domestic seed industry to enable it to produce seeds of high yielding varieties and hybrid seeds at a faster pace to meet the challenges of domestic requirements.

  (i) Seed Industry will be provided with a congenial and liberalized climate for increasing seed production and marketing, both domestic and international.

 (ii) Membership to International Organisations and Seed Associations like ISTA, OECD, UPOV, ASSINSEL, WIPO, at the National level or at the level of individual seed producing agencies, will be encouraged.

(iii) Emphasis will be given to improving the quality of seed produced and special efforts will be directed towards improving the quality of farmers' saved seeds.

(iv) Financial support for capital investment, working capital and infrastructure strengthening will be facilitated through NABARD/Commercial Banks/Cooperative Banks.

(v) Tax rebate/concessions will be considered on the expenditure incurred on in-house research and development of new varieties and other seed related research aspects. In order to develop a competitive seed market, the States will be encouraged to remove unnecessary local taxation on sales of seeds.

(vi) To encourage seed production in non-traditional areas including backward areas, special incentives such as transport subsidy will be provided to seed producing agencies operating in these marginalised areas.

(vii) Reduction of import duty will be considered on machines and equipment used for seed production and processing which are otherwise not manufactured in the country.

Q13. Describe strengthening of monitoring system in relation to National Seeds Policy 2002.

Ans. The Department of Agriculture and Cooperation (DAC) will supervise the overall implementation and monitoring of the National Seeds Policy.

(i) The physical infrastructure in terms of office automation, communication facilities, *etc.*, in DAC will be augmented in a time bound manner.

(ii) The technical capacity of DAC need to be augmented and strengthened to undertake the additional work relating to implementation of National Seeds Policy, implementation of PVP and FR Bill, Seeds Act, Import and Export of Seeds, *etc.*

(iii) Capacity building, including National and International training and participation in Seminars/Workshops will be organized for concerned officials.

Chapter 9

Indian Seeds Act 2004

Q1. What is Indian Seed Act 2004?

Ans. The Indian Seeds Bill 2004 hereinafter called Indian Seeds Act 2004 was approved by the government of India in 2004 and came into force with effect from. January 2005. The main objective of the Seeds Bill, 2004 is to ensure availability of quality seeds to farmers. It replaces the Seeds Act, 1966.

Q2. What the scope of Indian seeds Act 2004?

Ans. This Act has wider coverage than seeds act of 1966. It includes agriculture, horticulture, forestry, plantation crops, medicinal and aromatic plants. The seeds of all these crops are covered by the new act.

Q3. What category of seeds permitted by the seeds Act 2004?

Ans. This act permits sale of certified seed only The name of variety, physical purity and germination percentage have to be indicated on the seed container or bag.

Q4. Is license compulsory?

Ans. The seed dealers, sellers and growers should have license which is compulsory.

Q5. Is registration compulsory?

Ans. Registration for all kinds and varieties of seed is compulsory. No producer can grow seeds unless he is registered. It is compulsory for all cultivated varieties of field crops, vegetable crops, fruit crops, medicinal and aromatic plants, forest species and plantation crops. Every seed producer and dealer, and horticulture nursery has to be registered with the state government.

Q6. What are the Criteria for Registration?

Ans. The released and notified varieties can be registered for seed multiplication. There are four criteria for protection of varieties under PPV and FR Act. These are novelty, distinctiveness, uniformity and stability.

Q7. Period of Registration.

Ans. The Indian Seeds Act 2004 has adopted UPOV Act 1978. According to this Act, the period of protection is 15 years for annual and biennial crops and 18 years for perennial plants such as trees and vines.

Q8. How the Certification is done?

Ans. The Act permits self certification of seeds by accredited agencies and allows the central government to recognize certification by foreign seed certification agencies. Seed producers are permitted to self certify the performance of their seed under certain conditions. The seed companies need to provide the results of multiplication trials before registration.

Q9. What is the procedure of claiming Compensation?

Ans. The disputes about the quality of seeds have to be settled in the consumer court. Any loss in the production due to poor seed quality is claimed in the consumer court.

Q10. What are the Farmers' Privilege in the seeds Act 2004?

Ans. The Seeds Act 2004 allows farmers' privilege. The Bill does not restrict the farmer's right to use or sell his farm seeds and planting material, provided he does not sell them under a brand name. All seeds and planting material sold by farmers will have to conform to the minimum standards of germination, physical purity and genetic purity applicable to registered seeds.

Q11. What are the Penalties for violation of the Act 2004?

Ans. Any person who contravenes any provisions of the Act or imports, sells or stocks seeds deemed to be misbranded or not registered, can be punishable by a fine between Rs 5,000 and Rs 25,000. The penalty for giving false information is a prison term up to six months and/or a fine up to Rs 50,000. In case of companies, the person(s) in charge of conduct of the business of the company will be held accountable.

Q12. What are the technologies which have been Excluded in the seeds Act 2004?

Ans. Those techniques which are harmful to animals and the environment such as terminator technology and traitor technology are excluded and not allowed by the seed act.

Q13. What Research Exemption have been provided in the seeds Act 2004?

Ans. This act allows use of protected material by breeders for development of new varieties.

Q14. What is the structure of Central seed Committee in the seeds Act 2004?

Ans. the Seed Act 2004, the central seed committee consists of Chairman [Secretary DARE], Agricultural Commissioner, DDG horticulture [ICAR], Joint Secretary Seeds [DAC], representatives from DBT, Ministry of Environment and Forest, Secretary Agriculture [from five States], Director [SSCA] from one State and MD-SSC from one State. In addition, two representatives each of farmers and seed industry are nominated.

Q15. What are the differences between Seeds Act 1966 and Seeds Act 2004?

Ans. There are several differences between seeds act 1966 and seeds Act 2004 which are presented in Table 15-1.

TABLE 15-1: Differences between Seed Act 1966 and Seed Act 2004

Sl.No.	Particulars	Seeds Act 1966	Seeds Act 2004
1.	Crop covered	Agriculture and horticulture	Agriculture, horticulture, forestry, plantation crops, medicinal and aromatic plants
2	Registration of transgenic varieties	No provision	Special provision
3	Registration with PPVFR authority	Not required	Required
4	Period of protection	Not defined	Defined
5	Penalty for violation of Act	Rs 100-1000/and 6 months imprisonment	Rs 500-50,000/and six months imprisonment
6	Self Certification	Not permitted	Permitted
7	Seed multiplication of land Races	Permitted	Not permitted
8	Representatives in central seed committee	From all States	From five States only
9	Involvement of Private Sector	No	Yes

Q16. What are the similarities between Seeds Act 1966 and Seeds Act 2004?

Ans. There are several similarities between seeds act 1966 and seeds Act 2004 which are presented in Table 15-2.

TABLE 15-2: Similarities between Seed Act 1966 and Seed Act 2004

Sl.No.	Particulars	Seeds Act 1966	Seeds Act 2004
1	Category of seed permitted for sale	Certified Seed	Certified Seed
2	License	Compulsory	Compulsory
3	Certification	Compulsory	Compulsory
4	Farmers' exemptions	Available	Available
5	Purity and germination standards	Adopted	Adopted
6	Compensation of loss by use of certified seed	Through consumer court	Through consumer court
7.	Requirements for seed multiplication	Released and notified varieties	All seeds for sale must be registered

Q17. What are the differences between Seeds Act 2004 and PPV and FR Act 2001?

Ans. There are several differences between Seed Act 2004 and Protection of Plant Variety and Farmers' Rights Act 2001 which are presented in Table 15-3

TABLE 15-3: Differences between Seeds Act 2004 and Protection of Plant Varieties and Farmers' Rights Act 2001

Sl.No.	Particulars	Seeds Act 2004	PPV and FR Act 2001
1	Record Maintained	National Register of Seeds	National register of plant varieties
2	Compensation	Through consumer court	Through PPV and FR Authority
3	Disclosure of variety's parentage	Not required	Required
4	Variety's performance to be declared by	Seed producer, distributor or vendor	Breeder of the variety
5	Penalty for violation of the Act	Rs 5000-50,000/	Rs 50,000-20lakh
6	Imprisonment for violation of Act	Up to six months	From 2-3 years

Q18. What are the similarities between Seeds Act 2004 and PPV and FR Act 2001?

Ans. There are several similarities between Seed Act 2004 and Protection of Plant Variety and Farmers' Rights Act 2001 which are presented in Table 15-4

TABLE 15-4: Similarities between Seeds Act 2004 and Protection of Plant Varieties and Farmers' Rights Act 2001

Sl.No.	Particulars	Seeds Act 2004	PPV and FR Act 2001
1	Registration	Compulsory	Compulsory
2	Period of registration	15 years for annual and biennial crops and 18 years for perennials	15 years for annual and biennial crops, 18years for perennials
3.	Renewal of registration	Permitted for similar period	Permitted for similar period
4	Farmers Rights	Allowed	Allowed
5	Plant breeders' Rights	Available	Available
6	Disclosure of variety's performance	Compulsory	Compulsory
7	Sale of branded seeds by farmers	Not permitted	Not permitted

Q19. What are the Merits of Seed Act 2004?

Ans. There are several merits of the seeds Act 2004 which are briefly discussed as follows:

(i) **Wider Coverage.** This act has wider coverage [agriculture, horticulture, forestry, plantation crops, medicinal and aromatic plants] than Seed Act of 1966.

(ii) **Easy Certification.** This act has made the process of certification very simple means allowed self certification. But it should not be misused.

(iii) **Bans Harmful Technologies.** The act did not permit registration of harmful technologies such as terminator technology and traitor technology.

(iv) **Permits Farmers' Rights.** It permits farmers' rights. The farmers need not to register their varieties.

(v) **Ensures Quality Seed.** This act ensures availability of good quality seeds to farmers by imposing heavy penalties on the sale of spurious seeds.

(vi) **Registration for long Duration.** This act provides registration for 15 years for annual and biennial crops and 18 years for perennial crops which is sufficiently long period. Moreover, the registration is renewable for another term of similar duration.

(vii) **Central Seed Committee.** In the Seed Act 2004, the constitution of central seed committee is better than that of Seed Act 1966.

(viii) **Promotes Seed Industry.** The new seed act will lead to fast development of seed industry due to involvement of representatives of multinational seed companies or industries in the central seed committee.

  (ix) Increase in Productivity. The new act will help in increasing crop productivity due to availability of better quality seeds.

  (x) The present seed replacement rate is 20-25 per cent. The seed act will enhance seed replacement rate of various crops.

  (xi) The seed act will boost the export of seeds and encourage import of useful germplasm for use in crop improvement programmes.

  (xii) It will encourage investment in seed research and development.

Q20. What are the Demerits of Seed Act 2004?

Ans. There are some demerits of the seeds Act 2004 which are briefly discussed as follows:

(1) Although farmers are exempt from registering their seed varieties, the seeds have to conform to standards prescribed for commercial seeds. Farmers may find it difficult to adhere to the standards required of commercially sold seeds, because the seed saved and exchanged by farmers constitute above 80 per cent of the seeds planted

(2) It is not clear whether the compensation would include the value of the crop or only the cost of the seed.

(3) It is not clear whether a seed producer may sell seed which is registered by a different producer.

(4) The disclosure of parentage is not essential. It may lead to a situation where seeds can be registered without disclosing the parentage or origin of the seed.

(5) The Act does not have the provision of benefit sharing unlike the Convention on Biological Diversity and the PPVFR Act.

(6) The Seed Act does not provide for a mechanism to trace back a packet of seed to the dealer, processor and producer. This would make it difficult to trace back a defective lot, and rectify any deficiencies in the supply chain.

(7) Seed producers would be permitted to self-certify the performance of their seeds under certain conditions. This opens up the possibility of false declaration by seed companies. To prevent this, only government agencies should be allowed to conduct these trials and grant certification.

(8) Every horticultural nursery has to be registered with the state government and has to maintain records of layout plan, source of every planting material *etc.* Such nurseries in the unorganized sector may find it difficult to adhere to these conditions.

(9) The Seed Inspector has the power to enter and search as well as break open container or break open doors, without a warrant. This is against the normal procedure.

(10) The new seeds act will encourage cultivation of new varieties. Thus old varieties and land races will be replaced by new varieties on vast areas resulting in significant reduction in biodiversity. Reduction in biodiversity will invite danger of uniformity and lead to narrow genetic base and narrow adaptation.

(11) It will encourage multinational seed companies which may lead to monopoly of such companies. Seed companies may hike seed prices resulting in exploitation of farmers.

Varieties Related Legislations

Protection of Plant Varieties and Farmers' Rights Act 2001

Q1. What are Protection of Plant Varieties and farmers' Rights Act 2001?

Ans. The Protection of Plant Varieties and Farmers' Rights Act 2001 was approved by the government of India in 2004 but came into force with effect from. January 2005. The main objective of this act is to encourage development of new plant varieties and promote development of seed industry in India.

Q2. What are the objectives of Protection of Plant Varieties and farmers' Rights Act 2001?

Ans. There are four main objectives of enforcing the Protection of Plant varieties and Farmers' Rights Act 2001 as given below:

(i) To establish an effective system for protection of plant varieties, farmers' rights and plant breeders' rights

(ii) To recognize and protect farmers' rights for their contribution in conserving, improving and making available plant genetic resources for development of new plant varieties.

(iii) To stimulate investment for research and development both public and private sector for development of new plant varieties.

(iv) To promote the growth of seed industry in the country to ensure availability of high quality seeds to the farmers.

Q3. What is the Legislative Structure of PPV and FR Act 2001?

Ans. The Authority consists of a chair person called as authority and 15 persons as members. The members of the Authority include (i) Agricultural Commissioner-GOI, (ii) DDG Crop Sciences-ICAR, (iii) Joint Secretary

Seeds-GOI, (iv) Horticulture Commissioner-GOI, (v) Director, NBPGR; (vi) one member each [not below the rank of Joint Secretary] from DBT, (vii) Ministry of Environment and forests-GOI, (viii) Ministry of Law Justice and Company Affairs- GOI; (ix) one representative each [nominated by central government] from farmers, (x) tribal organization, (xi) seed industry, (xii) an agricultural university, (xiii) and women organization; and (xiv) two nominated representative of state government.

Q4. What is the Headquarters of PPV and FR Act Authority?

Ans. The headquarters of Protection of Plant Varieties and Farmers' Rights Authority is located in Pusa Campus [National Agricultural Science Complex], New Delhi.

Q5. What are the Plant Material Protected.by PPV and FR Act 2001?

Ans. Various types of crop cultivars can be protected under PPV and FR Act. The following breeding material can be protected.

    (i) All new varieties of self pollinated species such pure line and multi line varieties.

    (ii) New varieties of cross pollinated species such as open pollinated variety, synthetics and composites.

    (iii) New varieties of vegetatively/clonally propagated species.

    (iv) Single, three-way and double cross hybrids.

    (v) Inbred parental lines of hybrid varieties.

    (vi) Trees, fruits, ornamental plants and vines, *etc.*

Q6. What are the requirements for Registration. of new variety under PPV and FR Act 2001?

Ans. The PPV and Fr Act Authority establishes a Plant Varieties Registry, which maintains a National Register of Plant Varieties. Specifies details under which a variety may be registered such as a complete passport data of the parental lines from which a variety has been derived have to be furnished by the breeder. A new variety will be registered under the act if it conforms to the criteria of novelty, distinctiveness, uniformity and stability.

    (i) **Novelty.** It refers to newness of a variety. A variety which has not been grown for more than one year. Prior the application for registration can be protected.

    (ii) **Distinctiveness.** The variety should be clearly distinguishable fro previously available varieties at least for one characteristics.

    (iii) **Uniformity.** It refers to same type of population of a variety. In cross pollinated species, the range of variation should be within the tolerable limit.

(iv) **Stability.** It means that the variety should give stable performance after repeated reproduction or propagation. Moreover, a variety should have a designation.

For registration of a variety, an application has to be filed with the Registrar, PPV and FR Act. No fee is charged if a variety is registered by a farmer.

Q7. What the duration of Protection under PPV and FR Act 2001?

Ans. The PPV and FR Authority has accepted UPOV Act 1978. According to this Act, the period of protection is 15 years for annual and biennial crops and 18 years for perennial plants such as trees and vines. The registration is renewable after its expiry for a similar term of duration.

Q8. What Farmers' Rights are provided by PPV and FR Act 2001?

Ans. A farmer is entitled to save, use, sow, resow, exchange, share or sell his farm produce including seed of a variety protected under the Act in the same manner before this Act came into force. He cannot sell branded seed of a variety protected under the Act.

Q9. What Plant Breeders' Rights have been provided by PPV and FR Act 2001?

Ans. A certificate of registration for a variety issued by this act shall confer an exclusive right on the breeder or his successor, his agent or licensee, to produce, sell, market, distribute, import or export seed of his variety. The breeder has right to authorize any other person for production and marketing of his variety.

Q10. What are the technologies which have been Excluded in PPV and FR Act 2001?

Ans. Those techniques which are harmful to animals and the environment such as terminator technology and traitor technology are excluded and not allowed by the Protection of Plant Varieties and farmers' Rights Act 2001.

Q11. How the Compensation is claimed if a variety does not perform as declared?

Ans. If a breeder of a propagating material of a variety registered under the Act sells his product to a farmer, he has to disclose the expected performance under given conditions. If the propagating material fails to perform, the farmer can claim compensation in the prescribed manner before the Protection of Plant Varieties and Farmers' Rights Authority.

Q12. What are the Penalties for violation of the PPV and FR Act 2001?

Ans. Penalty for applying false denomination to a variety is imprisonment up to two years and/or a fine between Rs 50,000 and Rs five lakh. Penalty for falsely representing a variety as registered is imprisonment up to three years and/or a fine between Rs one lakh and Rs five lakh or both. Penalty for subsequent offence is imprisonment up to three years and/or a fine between Rs two lakh and Rs 20 lakh.

Q13. What are the similarities between Seed Act 2004 and PPV and FR Act 2001?

Ans. There are some similarities between Seed Act 2004 and Protection of Plant Variety and Farmers' Rights Act 2001 which are presented in Table 10-1.

TABLE 10-1: Similarities between Seeds Act 2004 and Protection of Plant Varieties and Farmers' Rights Act 2001

Sl.No.	Particulars	Seeds Act 2004	PPV and FR Act 2001
1	Registration	Compulsory	Compulsory
2	Period of registration	15 years for annual and biennial crops, 18years for perennials	15 years for annual and biennial crops, 18years for perennials
3.	Renewal of registration	Permitted for similar period	Permitted for similar period
4	Farmers Rights	Allowed	Allowed
5	Plant breeders' Rights	Available	Available
6	Disclosure of variety's performance	Compulsory	Compulsory
7	Sale of branded seeds by farmers	Not permitted	Not permitted
8	Involvement of Private Sector	Yes	Yes

Q14. What are the differences between Seed Act 2004 and PPV and FR Act 2001?

Ans. There are several differences between Seed Act 2004 and Protection of Plant Variety and Farmers' Rights Act 2001 which are presented in Table 10-2.

TABLE 10-2: Differences between Seeds Act 2004 and Protection of Plant Varieties and Farmers' Rights Act 2001

Sl.No.	Particulars	Seed Act 2004	PPV and FR Act 2001
1	Record maintained	National Register of Seeds	National register of plant varieties
2	Compensation	Through consumer court	Through PPV and FR Authority
3	Disclosure of variety's parentage	Not required	Required
4	Variety's performance to be declared by	Seed producer, distributor or vendor	Breeder of the variety
5	Penalty for violationof the Act	Rs 5000-50,000/	Rs 50,000-20lakh
6	Imprisonment for violation of the Act	Up to six months	From 2-3 years

Q15. What are the Merits of PPV and FR Act 2001?

Ans. There are several advantage of Protection of Plant varieties and Farmers' Rights Act 2001 which are briefly presented below:

(i) It provides an effective system for protection of plant varieties, farmers' rights and plant breeders' rights

(ii) It recognizes the contribution of farmers in conserving, improving and making available plant genetic resources for development of new plant varieties.

(iii) It provides rights to farmers to save, use, sow, resow, exchange, share or sell his farm produce including seed of a variety protected by PPV and FR Act.

(iv) It confers an exclusive right on the breeder or his successor, his agent or licensee, to produce, sell, market, distribute, import or export seed of his varietyThus it gives recognition to breeders for their contribution and encourage them to breed still better varieties.

(v) It encourages investment in plant breeding research and development both by public and private sectors for development of superior plant varieties. It will lead to fast development of seed industry and improvement in seed quality due to competitions among breeders.

(vi) It promotes the growth of seed industry in the country to ensure availability of high quality seeds to the farmers.

(vii) It involves representatives of farmers, women and seed industry in its legislative structure who will look to the interest of these sectors.

(viii) The Act has provision for compensation. If the seed gives poor performance than the declared by the breeder, farmer can claim compensation in the prescribed manner before the Protection of Plant Varieties and Farmers' Rights Authority.

(ix) The Act has provision for rewarding farmers for their role in conserving plant genetic resources. Such expenditure can be met out from the gene fund available with the PPV and FR Act Authority.

(x) The act has provision of benefit sharing. The farmers can get share from the benefits arising from cultivation of farmers varieties and knowledge.

(xi) The act has provision to impose penalties to curb unhealthy practices.

Q16. What are the Demerits of PPV and FR Act 2001?

Ans. There are some demerits of Protection of Plant varieties and Farmers' Rights Act 2001 which are briefly presented below:

(i) The PPV and FR Act will promote cultivation of new high yielding varieties which are highly uniform. This will result in reduction in genetic diversity leading to narrow genetic base and danger of uniformity.

(ii) This will encourage monopoly of some breeders or seed companies for genetic material with special traits

(iii) The holders of plant breeders rights will produce less seed than market demand and hike price to earn more profit. The high price can not be afforded by small and marginal farmers Thus all farmers can not take advantage of new varieties.

(iv) It will inhibit free exchange of seed material and encourage unlawful practice.

Chapter 11

Plant Breeders' Rights

Q1. What are Plant Breeders' Rights?

Ans. Legal rights granted to the breeder of a new variety of plant for its production and marketing are called plant breeders' rights. Plant Breeders Rights are granted to novel plant varieties that are distinctive, uniform, and stable (e. g, cultivars breed true-to-type for desired traits). The legal protection of a new plant variety is granted to the breeder or his successor. The effect of PBR is that prior authorization is required before the material can be used for commercial purposes.

Q2. What is the procedure for Registration of Plant Breeders' Rights?

Ans. Registration is essential to get legal rights. There is no protection without registration. The registration consists of following steps:

(i) Filing application in the prescribed form,

(ii) Payment of prescribed processing fee along with application. The registration fee varies from country to country. In USA, after registration annual renewal fee [300 Dollars] is payable.

(iii) Examination of the application by the experts, and

(iv Issue of protection certificate, if application if found to meet the desired requirements.

(v) Normally the registration takes about 2.5 years for most of the species. takes 2.5 years.

Q3. What is the Duration for protection of Plant Breeders' Rights?

Ans. The period of protection varies with plant species. The protection is provided for a definite period. The duration of protection for different plant species is as follows'.

   (i) For Trees and Vines: Initially for nine years and maximum for 18 years.

  (ii) For extant varieties: Initially six years and maximum for 15 years fro the date registration of the variety.

 (iii) In other cases: Initially six years and maximum for 15 years fro the date registration of the variety.

Thus breeders' rights allow plant breeders to get benefit of their variety for a limited period. When the PBR period is over, the variety reverts to the public domain and is available to every body. The limited duration of Plant Breeders' Rights ensure a balance between private and public interest.

Q4.   What is the Validity of PBRs?

Ans.   The protection of the Plant Breeders' Rights is valid only in the country where it has been registered. The protection in other countries can be obtained by filling separate application in each country.

Q5.   What are the Matters Covered by Plant Breeders' Rights?

Ans.   The protection right can be granted for varieties of all botanical genera and species. The variety should have a designation [name] as per the rule of International Code of Nomenclature.6.

Q6.   What are the Requirements for protection of Plant Breeders' Rights?

Ans.   There are four basic requirements for protection of a variety under PBR, *viz..* (i) novelty, (ii) distinctiveness, (iii) uniformity, and (iv) stability. See later.

Q7.   Is it possible to Transfer Plant Breeders' Rights?

Ans.   An authorized plant breeder has right to authorize other interested persons for commercial production and marketing of his variety on royalty basis.

Q8.   Who has Control over the Plant Breeders' Rights?

Ans.   The authorized plant breeder has right to prevent other from commercial production and marketing of his variety without permission.

Q9.   When does the Plant Breeders' Rights become effective?

Ans.   The Plant Breeders' Rights come into force immediately after registration of the variety.

Q10. To whom plant breeders rights are granted?

Ans.   BPR can be granted only to breeder or legal owner of a new variety.

Q11. What is protected by plant breeders' rights ?

Ans.   The plant breeders rights protect the variety but not the standard breeding procedures that are used to develop new variety.

Q12. What is the purpose of plant breeders' rights?

Ans. The main purpose of PBR is to encourage breeders to develop better plant varieties. The PBRs provide monopoly of new varieties to breeders for a limited period.

Q13. Can the same variety be patented and protected by plant breeders' rights?

Ans. Yes, the same variety can be patented and protected by PBR. The gene technology or gene sequences can be patented and the end product that is the plant variety can be protected under PBR.

Q14. What are the Requirements for protection of plant breeders' rights?

Ans. PBR cannot be granted to any variety. There are four basic requirements for protection of a variety under Plant Breeders' Rights. These are (i) novelty, (ii) distinctiveness, (iii) uniformity, and (iv) stability.

Q15. What do you mean by Novelty?

Ans. It refers to newness of a variety. The variety should be new and should not have been commercially cultivated for more than one year before granting protection under PBR. In other words, there should not be a prior sale for more than one year.

Q16. What is distinctiveness?

Ans. The new variety must be distinguishable in one or more morphological, quality or other character from previously available varieties. When a variety differs atleast in one characteristic from varieties of common knowledge, it is known as distinct.

Q17. Define Uniformity.

Ans. The variety should be uniform. In other words, all plants in a variety should look alike.

Q18. What is stability?

Ans. The should give stable performance in different generations. All these attributes are determined by grow out test.

Q19. What are the Rights Provided to the breeder of a new variety?

Ans. The *Plant Breeders' Rights Act* provides plant breeders the exclusive right to produce and sell new plant varieties which they have developed. In other words, It provides exclusive rights to the breeder for commercial production and marketing of his variety. The important Plant Breeders' Rights are as follows:

(i) **Rights for commercial Seed production**: PBR provide legal right to the breeder of a variety for large scale seed production. This he can do either on his own farm or on the farms of authorized farmers on payment basis.

(ii) **Rights for Marketing**: The breeder or owner of a variety has exclusive rights to regulate marketing of his variety.

(iii) **Rights to export and import**: The breeder or owner of a variety has full rights to regulate export and import of his or her variety.

(iv) **Rights of Authorization.**: The breeder or owner of a protected variety has rights to authorize other interested persons for commercial production, and marketing, export and import of his variety. However, prior authorization of the breeder or owner of a variety is required for such purpose.

(v) Rights to prevent Infringement: The breeder or owner of a variety has rights to prevent others from unauthorized commercial production and marketing of his variety.

Q20. What is Infringement of plant breeders' rights?

Ans. Unauthorized production and marketing of a registered variety by other person amounts to infringement. The owner has the right to take legal action against the infringer and claim damages. The PBR Authority may initiate legal action against a person who is involved in the infringement of a protected plant variety. The Authority can recover both damages and profits from such person. The Plant Breeders' Act [PBRA] provides for heavy penalties against infringement of the breeders' right. In USA it 55,00 Dollars for individuals and 275000 Dollars for companies.

Q21. What are exemptions of Plant Breeders' Rights Act?

Ans. The Plant Breeders' Rights Act provides two exemptions, *viz..* (i) breeders' exemptions and (ii) farmers' exemptions.

Q22. What do you mean by Breeders' Exemptions?

Ans. The legal rights that are provided to plant breeders to use protected material for further research refer to breeders' exemptions. Breeders exemptions are also called research exemptions or breeders' privilege. The UPOV Act 1978 provides Breeders' exemptions. However, the Act 1991 has curtailed Breeders' Exemptions.

Q23. What are Farmers' Exemptions?

Ans. The legal rights that are provided to farmers to save, use, exchange, share or sell his farm produce of a protected variety are known as farmers' exemptions. Farmers' exemptions are also called Farmers' Rights or Farmers privilege. Here the sale is restricted to non-commercial sale. The UPOV Act 1978 provided farmers Exemptions, however, these exemptions were curtailed by UPOV Act 1991.Because of these reasons, UPOV Act is not accepted by many countries.

Q24. What are Advantages of plant breeders' rights?

Ans. There are several advantages of plant breeders' rights which are listed below:

   (i) Breeders get benefit of their variety

   (ii) PBR help in faster development of seed industry.

   (iii) PBR lead to improvement in quality because of competition

   (iv) PBR are useful in procurement of good material on payment basis.

   (v) PBR help in enrichment of genetic resources.

Q25. What are Disadvantages of plant breeders rights?

Ans. There are some disadvantages of plant breeders' rights which are listed below

   (i) It will promote monopoly

   (ii) It will encourage unhealthy practices

   (iii) It may lead to increase in prices.

   (iv) There will be reduction in genetic variability

   (v) There will be compulsion to purchase fresh seed every year.

Chapter 12

Farmers' Rights

Q1. What are Farmers' Rights?

Ans. Farmers' rights refer to the rights arising from the past, present and future contributions of farmers in conserving, improving and making available plant or animal genetic resources, particularly those in the center of origin/ diversity. In other words, the legal rights provided to farmers to save, use, sow, replant, exchange, share or sell his farm produce including seed of a variety protected under Plant Variety Protection Act refer to Farmers Rights. The purpose of these rights is to "ensure full benefits to farmers and support the continuation of their contributions.

The FAO Conference held in Rome from 11-29 November, 1989 endorsed the concept of Farmers Rights with a view to:

(i) Ensuring global recognition of the need for conservation and the availability of sufficient funds for these purposes;

(ii) Assisting farmers and farming communities throughout the world, especially those in areas of original diversity of plant genetic resources, in the protection and conservation of their PGR and of the natural biosphere; and,

(iii) Allowing the full participation of farmers, their communities and countries in the benefits derived, at present and in the future, from the improved use of PGR.

Q2. Is Registration required for a farmers' Variety?

Ans. A farmer who has bred or developed a new variety shall be entitled for registration and protection of his variety like a breeder.

Q3. What is the procedure of registration of farmers' variety?

Ans. The registration process consists of the following steps:

(i) **Filing of application.** The application is to be filed in the prescribed form and submitted in the office of the registrar of Plant Variety Protection and farmers' Rights Act. The application should contain all the desired documents and information

(ii) **Examination of the application.** The application is examined by experts in the registrar office. If found correct, it is advertised for opposition. Three months time is given for filing opposition.

(iii **Registration.** If there is no opposition, and the registrar and Authority are satisfied, the variety is registered and the registration certificate is issued to the breeder, farmer or other owner of the variety.

Q4. What are the Requirements for registration of farmers' variety?

Ans. The farmers; variety shall be entitled for registration if it meets the following requirements.

(i) It should have a denomination [name] assigned by the applicant

(ii) An affidavit that the variety does not contain terminator gene.

(iii) A complete passport data of parental lines from which the variety has been derived

(iv) The contribution of any farmer, village community, institution or organization in the development of the variety.

(v) The source of genetic material used in developing the variety and the parental material used in developing new variety was lawfully acquired.

(vi) A brief description of the variety about its novelty, distinctiveness, uniformity and stability.

(vii) A proof of the right to make the application

Q5. How much fee is charged from farmers for registration of a variety?

Ans. In India, a farmer or group of farmers or village community shall not be liable to pay any fee in any proceeding before the Authority or Registrar or the Tribunal or the High Court under Plant Variety Protection and Farmers' Rights Act or the rules made there under.

Q6. What are the Rights granted to farmers?

Ans. The Farmers' Rights Act provide following rights to the farmers in India even for the variety protected under PVP and FR Act.

(i) Rights to save, use, sow or replant seeds of his farm produce.

(ii) Rights to exchange or share his farm produce, and

(iii) Rights to sell seeds of his farm produce

However, the farmer cannot sell the seeds or planting material under a brand name.

Q7. What are Rights of Communities in India?

Ans. In India, any of the following can file application for registration on behalf of any village or community.

(i) Any person or group of persons [whether actively engaged in farming or not], and

(ii) Any governmental or non-governmental organization

After completing all formalities, if the Authority is satisfied, the variety can be registered. If due credit has not been given to the farmers or communities who have helped in the development of the variety, the breeder has to pay compensation to the concerned party. Or part of compensation will be deposited in the gene fund.

Q8. Is Authorization of Farmers' Variety possible?

Ans. The authorization for an essentially derived variety from farmers' variety cannot be given by the breeder of such farmers; variety except with the consent of the farmers or group of farmers or community of farmers who have made contribution in the preservation or development of such variety.

Q9. What are the Materials Covered by Farmers' Rights?

Ans. Farmers' Rights include new varieties of all self pollinated, cross pollinated and asexually propagated varieties.

Q10. What is the Duration for Farmers' variety?

Ans. The protection is provided for a definite period. The duration of protection for different plant species is as follows'

(i) For Trees and Vines: Initially for nine years and maximum for 18 years.

(ii) For extant varieties: Initially six years and maximum for 15 years fro the date registration of the variety.

(iii) In other cases: Initially six years and maximum for 15 years fro the date registration of the variety.

Q11. What is the Validity of protection of Farmers' Rights?

Ans. The protection is valid only in the country where the variety has been registered.

Q12. What is Benefit Sharing ?

Ans. If the variety has been developed by a particular farmer, he can get benefits of his variety. In case of community rights, the sharing of benefits becomes difficult and the money is deposited in the gene fund.

Q13. What is Gene Fund?

Ans. The central government may constitute a Fund to be called National Gene Fund. The money earned from following will be deposited in this fund;

(i) The benefit sharing received fro the breeder of a variety

(ii) The annual payable to the Authority in the form of royalty.

(iii) The amount of compensation received by the Authority.

(iv) Contribution received from amy National or International Organization and other sources.

The gene fund may be utilized for benefit sharing, rewards and other expenditure related to Plant Variety Protection and Farmers' Rights Act.

Q14. Who will be entitled for Rewards?

Ans. A farmer who is engaged in the conservation of genetic resources of land races and wild relatives of economic plants and their improvement through selection shall be entitled for recognition and reward. The expenses toward rewards will be met out from the National Gene Fund

Q15. What do you mean by Acknowledgement by Breeders?

Ans. It is compulsory for the breeder of a variety to reveal the origin of the parental lines plant genetic resources used in the development of a variety. It will help in deciding the share of different persons, *viz..*, farmers and other communities in benefits earned by breeder of such variety.

Q16. Is Infringement is exempt for Indian farmers?

Ans. If a farmer was not aware of the PVP and FR Act at the time of using any protected variety, he will be exempt from the infringement and will be asked not to do such act in future. This right has been provided to Indian farmers by PVP and FR Act 2001.

Q17. What the purpose of farmers' rights?

Ans. Farmers' Rights, as endorsed by FAO in 1989, recognize that farmers and rural communities have contributed and continue to contribute to creation, conservation, exchange and enhancement of genetic resources and they should be recognized and strengthened in their work.

Q18. What the significance of farmers' rights?

Ans. The important points about nature and scope of Farmers' Rights are briefly presented as follows:

(i) **Ancient Rights.** Farmers' Rights have a deep historic character. They existed since humans created agriculture to serve their needs, and played vital role in conservation of biodiversity. Farmers are the guardians of these genetic resources, which support the evolution of species. Farmers are the inheritors of the skills and knowledge of the generations that have created this biological wealth. For these contributions, it is necessary to recognize Farmers' Rights.

(ii) **Rights Covered**. Farmers' Rights include the right over resources and associated knowledge, united indivisibly, and mean the acceptance of traditional knowledge, respect for cultures and recognition that these are the basis of the creation of knowledge

(iii) **Property Rights**. The farmers deserve right to control, the right to decide the future of genetic resources, the rights to define the legal framework of property rights of these resources.

(iv) **Collective Nature**. Farmers' Rights are of an eminently collective nature and therefore are protected by sui generic rights.

(v) **Application**. These rights should have a national application, respecting the sovereignty of each country, to establish local laws based on these principles

(vi) **Rights to Means**. It includes territorial rights, right to land, right to water and air and rights to conserve biodiversity and achieve food security.

(vii) **Active Participation**. The farmers deserve right to participation in the definition, elaboration and execution of policies and programmes linked to genetic resources

(viii) **Rights to Technology**. The farmers deserve right to appropriate technology as well as participation in the design and management of research programmes.

(ix) **Rights to Benefit Sharing**. The farmers deserve right to define the control and handling of benefits derived from the use, conservation and management of these resources.

(x) **Rights on Genetic Resources**. The farmers deserve right to use, choose, store and freely exchange of genetic resources.

(xi) **Rights for Sustainable Agriculture**. The farmers deserve right to develop models of sustainable agriculture that protects the biodiversity and to influence the policies that support it.

Q19. What is the Role of Farmers Related to PGR?

Ans. Farmers are both men and women who have domesticated, developed, conserved and made available plant genetic and other natural resources (land; water; vegetation; animal, bird and fish life) to which they have access in order to obtain a livelihood and ensure the well-being of their family through the provision of basic requirements such as food, fuel and water and income from the land.

Farmers have played key role in conservation of Plant Genetic Resource for thousands of years. The role of farmers in relation to PGR is briefly presented below:

(i) Farmers both men and women have conserved plant genetic resources over the past thousands of years.

(ii) Farmers are rightly called as on site managers of plant genetic resources

(iii) Farmers have in depth knowledge of the land races and wild relatives of cultivated plant species.

(iv) All modern plant varieties contain only those genes that have originated from farmers' traditional varieties (landraces) or wild crop relatives

(v) The women and men have historically provided the germplasm upon which scientific plant breeding is based.

(vi) Generally, germplasm is provided by small-scale subsistence farmers to plant breeders whose varieties are then adapted/developed for medium to large scale commercial farmers (Spillane, 1996).

Q20. What is the role of women related to PGR?

Ans. Women's special knowledge of the value and diverse uses of plants for nutrition, health and income has important implications for the conservation of plant genetic resources.

(i) The women have played significant role not only in terms of labour and skills, but also their decision-making about how natural resources are used to satisfy the multiple needs of rural households.

(ii) In much of the developing world, the on-farm and *in situ* conservation and use of plant genetic resources begins with women. As farmers, they are responsible for growing crop plants and collecting seed for next year sowing.

(iii) It is estimated that 90 per cent of the planting materials used in developing countries is produced by the farmers themselves, *i.e.* from saved seed. The selection of wild genetic resources for home planting, of seeds for conserving for next year's planting and of asexually propagated material is usually carried out by women. This selection is a sophisticated process that takes into account many different criteria such as taste, color, palatability and texture; resistance to pests and diseases; adaptation to soil and agro-climatic conditions and so on. The selection criteria are built up over years of experience.

Q21. What are the similarities between Plant Breeders' rights and farmers' Rights?

Ans. Plant Breeders' Rights and Farmers' Rights have some similarities which are presented in Table 12-1.

TABLE 12-1: Similarities between Plant Breeders' Rights and Farmers Rights

Sl.No.	Particulars	Plant Breeders'Rights	Farmers'Rights
1.	Registration	Essential	Essential
2.	Duration		
	Trees and vines	18 years	18 years
	Extant varieties *etc.*	15 years	15 years
3.	Enforcement	After Registration	After Registration
4.	Requirements	Novelty, Distinctiveness, uniformity and stability	Same as in PBR
5.	Validity	In country of registration	In country of registration

Q22. What are the differences between Plant Breeders' rights and farmers' Rights ?

Ans. Plant Breeders' Rights and Farmers' Rights have some differences which are presented in Table 12-2.

TABLE 12-2: Differences between Plant Breeders' Rights and Farmers Rights

Sl.No.	Particulars	Plant Breeders' Rights	Farmers' Rights
1.	Registration Fee	Charged	Not charged
2.	Annual renewal fee	Charged	Not charged
3.	Penalty for infringement	Charged	Not charged
4.	Community Rights	Not covered	Covered
5.	Authorization	Easy	Not permitted

Q23. Discuss the Future Outlook Related to PGR.

Ans. Plant Genetic Resources are the base materials for modern plant breeding programmes Hence their proper conservation and sustainable use are essential. In future, the following points should be given due importance for proper conservation and utilization of plant genetic resources.

(i) **Assistance to Farmers.** Farmers in all regions of the world should be assisted in their conservation of natural resources and more specifically the selection, conservation, improvement and sustainable use of PGR. Non Governmental Organizations should assist in the entire process as they are close to the communities and have the capacities to provide long-term support. The establishment of strong village level institutions with adequate representation of both women and men farmers can be key in promoting successful programmes.

(ii) **Incentives.** Appropriate incentives should be developed, with the active participation of those concerned, to convince communities, industry and governments of the benefits of conservation and development efforts and to provide financial resources, legal support, and technical capability to develop agricultural systems that are suitable for local agro-ecological conditions and, where possible, based on locally available resources. Incentives should be used to enable farmers to divert land, labour and capital towards conserving biological resources and to facilitate the participation of certain groups in activities that promote the sustainable use of those resources

(iii) **Participation of Women Farmers.** The participation of women farmers and the recognition of their contributions in the design and elaboration of such policies and programmes at all levels is essential. Women farmers should be included in decision making processes at community and regional levels. Farmers, communities and countries, and men and women at all levels should participate in the design and implementation of programmes policies and agreements related to PGR

(iv) **Self Decision.** Those farmers who develop breeding materials should be empowered to decide, at their discretion, if the materials will be made available to other (*i.e.* external) users.

(v) **Equal Rights to Women.** There should be equal rights for conservation and participation in PGR related programmes and policies, which is often lacking in developing countries. There should not be discrimination of gender. In the past, women farmers could get little benefit from support in terms of information and technology, inputs and credit, extension and research.

(vi) **Participatory Plant Breeding**. Participatory research and plant breeding should be promoted with different socio-economic groups in rural communities in order to improve the productivity of local varieties and to promote marketing of special products derived from these materials.

(vii) **Use of Land Races**. Modern plant breeding has concentrated on developing a small number of varieties that are widely adapted to many environments. Land races should also be used for developing varieties suitable for diverse ecological conditions.

(viii) **International Funds**. There is need for the creation of an Intentional Fund to support *in situ* conservation and to benefit rural men and women who produce and conserve PGR. In general farmers are interested in conserving resources but lack of financial support acts as a barrier in their way.

(ix) **Integrated Programs.** PGR conservation should be integrated into other community development programmes It will promote and employment in activities related to PGR building up of local institutions, capacity building) and collecting and dissemination;, information on the status and trends of PGR.

(x) **Benefit Sharing.** There should be equitable sharing of benefits arising from the utilization of PGR, knowledge, innovations and practices.

Q24. What are the advantages of Farmers' Rights?

Ans. There are several advantages of farmers' rights which are listed below:

    (i) Provide recognition to farmers, communities or village for their role in conserving Plant Genetic Resources.

    (ii) Provide respect to farmers

    (iii) Provide opportunities to farmers to participate in the policies, projects and programmes related to plant Genetic Resources [PGR].

    (iv) Lead to benefit sharing when farmers material is used in developing new variety by breeders.

    (v) Farmers permission is required to use their plant material

    (vi) Provide rewards to those farmers who are actively engaged in conserving Plant Genetic Resources.

Q25. What are the Limitations of Farmers' Rights?

Ans. There are some limitations of farmers' rights which are listed below:

    (i) Mostly farmers' rights are of collective nature

    (ii) The benefit sharing is difficult

    (iii) Farmers do not get any financial support from government for the conservation of Plant Genetic Resources.

    (iv) Farmers' Rights are of territorial nature.

Biodiversity Legislations

Chapter 13

Indian Biodiversity Legislation

Q1. What biodiversity?

Ans. The biodiversity refers to the variety of life in all forms, levels and combinations. The term biodiversity includes genetic diversity, species diversity, and ecosystem diversity. In other words, the variety of life forms – the different plants, animals and micro organisms, the genes they contain and ecosystems they may form is referred to as biodiversity. It is usually considered at three levels, *viz..*, genetic diversity, species diversity and ecosystem diversity. Crop biodiversity refers to the variety of genes and genotypes found in crop plants. Biodiversity is a modern term which simply means " the variety of life on earth". This variety of life can be measured in several ways.

Q2. What are different levels of biodiversity?

Ans. There are three levels of biodiversity, *viz..*, genetic diversity, species diversity, and ecological diversity. These levels are all interrelated yet distinct enough that they can be studied as three separate components.

Q3. Define Genetic Diversity.

Ans. **Genetic** diversity refers to the variation between individuals of the same species. This includes genetic variation between individuals in a single population, as well as variations between different populations of the same species. Genetic differences can now be measured using increasingly sophisticated techniques. These differences are the raw material of evolution. Genetic diversity is the variety present at the level of genes. Genes are made of DNA which determines expression of a trait in an individual. This level of diversity can differ by alleles (different variants of the same gene, such as yellow and white flower colour), by entire genes or by several genes.

Genetic diversity can be measured at many different levels such as population, species and community. The genetic diversity is important at each of these levels. The main points related to genetic diversity are listed below:

(i) It represents variation between individuals of the same species

(ii) It represents the variation at gene level

(iii) It can be measured at population, species and community levels.

(iv) Sophisticated techniques are required for its measurement.

(v) It represents the raw material for evolution and adaptation.

(vi) It is effected by the environment and competition with other species

Q4. What is the relationship between Genetic diversity and Adaptation?

Ans. The amount of diversity at the genetic level is important because it represents the raw material for evolution and adaptation. More genetic diversity in a species or population means a greater ability for some of the individuals in it to adapt to changes in the environment. Less diversity leads to uniformity resulting in poor adaptation For example, modern varieties of wheat and rice are high yielding and highly uniform. We can harvest a good crop from such varieties, but can be a problem when a disease or parasite attacks the field, as every plant in the field will be susceptible having the identical genetic constitution. Thus monocultures are unable to deal well with changing conditions.

Q5. What the effect of environment on Genetic Diversity?

Ans. Within species, genetic diversity often increases with environmental variability, which can be expected. If the environment often changes, different genes will have an advantage at different times or places. In this situation genetic diversity remains high because many genes are in the population at any given time. If the environment didn't change, then the small number of genes that had an advantage in that unchanging environment would spread at the cost of the others, causing a drop in genetic diversity.

Q6. What is the relationship between Genetic Diversity and Communities?

Ans. In communities, it can increase with the diversity of species. How much it increases depends not only on the number of species, but also on how closely related the species are. Species that are closely related (*e.g.* two species of wheat) have similar genetic structures and makeup and therefore do not contribute much additional genetic diversity. These closely related species will contribute to genetic diversity in the community less than more remotely-related species (*e.g.* a wheat and a barley) would.

An increase in species diversity can also affect the genetic diversity. If there are many species, the genetic diversity at that level will be larger than when there are fewer species. On the other hand, genetic diversity within each species can decrease. This can happen if the large number of species means so much competition that each species must be extremely specialized which will lead to little genetic diversity within any of the species.

Q7. What is Species Diversity?

Ans. **Species** diversity refers to the variety of species in a given region or area. This can either be determined by counting the number of different species present, or by determining taxonomic diversity. Taxonomic diversity is more precise and considers the relationship of species to each other. For example, an area containing three species of wheat and two species of barley, is more diverse than an area containing five species of wheat, even though they both contain the same number of species. High species biodiversity is not always necessarily a good thing. For example, a habitat may have high species biodiversity because many common and widespread species are invading it at the expense of species restricted to that habitat.

The species level crop biodiversity is the easiest to work on both from practical and theoretical point of view. Species are relatively easy to identify by eye in the field, whereas genetic diversity requires laboratories, time and resources to identify.

Species are well known and are distinct units of diversity. Each species can be considered to have a particular "role" in the ecosystem, so the addition or loss of single species may have consequences for the system as a whole The main points related to species diversity are listed below:

(i) It represents variety of species in a given region or area.

(ii) It identification is easy by naked eye in the field.

(iii) It can be determined simple by counting the number of different species.

(iv) Species are distinct units of diversity.

Q8. What is Ecosystem Diversity?

Ans. **Ecosystem** diversity refers to communities of plants and animals, together with the physical characteristics of their environment (*e.g.* geology, soil and climate). Ecosystem diversity is more difficult to measure because there are rarely clear cut boundaries between different ecosystems and they grade into one another. However, if consistent criteria are chosen to define the limits of an ecosystem, then their number and distribution can also be measured.

Ecosystem-level diversity deals with species distributions and community patterns, the role and function of key species, and combines species functions and interactions. This is the least-understood level of the three types of biodiversity due to the complexity of the interactions.

One of the difficulties in examining communities is that the transitions between them are usually not very sharp. A lake may have a very sharp boundary between it and the deciduous forest it is in, but the deciduous forest will shift much more gradually to grasslands or to a coniferous forest. This lack of sharp boundaries is known as "open communities" (as opposed to "closed communities," which would have sudden transitions) and makes studying ecosystems difficult. The main points related to ecosystem biodiversity are given below:

(i) It represents communities of plants and animals together with physical structures of their environment.

(ii) The measurement of ecosystem diversity is more difficult than genetic diversity and species diversity.

(iii) The assessment of ecosystem diversity needs many complex measurements to be taken over a long period of time

(iv) It is a time consuming task and completion of such work requires adequate staff and financial support.

Q9. What are the differences between genetic diversity and species diversity?

Ans. There are several differences between genetic diversity and species diversity which are presented in Table 13-1.

TABLE 13-1: Comparison of Genetic Biodiversity and Species Biodiversity

Sl.No.	Particulars	Genetic Diversity	Species Diversity
1	Represents	Variation between individuals of same species.	Variety of species in a given region
2	Identification	Difficult	Easy
3	Assessment	Requires sophisticated laboratory techniques	Can be assessed by counting the species
4	Conservation	Possible in gene banks	Possible in gene banks
5	Environmental effects	Present	Present

Q10. What are Advantages of Crop Biodiversity?

Ans. There are several advantages of crop biodiversity. Some important advantages or benefits of crop biodiversity are briefly presented as follows:

The crop biodiversity provides:

(i) Protection from biotic stresses such as diseases and insects.

(ii) Protection from abiotic stresses such as drought, soil salinity, soil alkalinity, soil acidity, cold temperature, heat *etc.*

(iii) Choice of quality for food and other items

(iv) Choice of items for food, fibre, oil, fuel, fodder to animals, timber and medicines.

(v) Wider adaptation to environmental changes

(vi) Broad genetic base to the populations

Q11. What are the reasons for Loss of crop biodiversity?

Ans. There are two main reasons of depletion of biodiversity in crop plants, *viz..*, extinction and genetic erosion. Extinction refers to permanent loss of a crop species due to various reasons. Genetic erosion refers to the gradual reduction in genetic variability, in population of a species, due to elimination of various genotypes. There are five main reasons of genetic erosion as follows:

   (i) Replacement of Land Races by Modern Cultivars.

  (ii) Industrial agriculture

 (iii) Farming into wild habitat.

  (iv) Clean Cultivation.

  (vi) Developmental Activities such as development of national high ways, towns, cities, airports, seaports *etc.*

Q12. What the need for Conservation of Biodiversity?

Ans. Conservation refers to protection of genetic diversity of crop plants from genetic erosion. The loss of biodiversity poses a serious threat to agriculture and the livelihoods of millions of people. Conserving biodiversity and using it wisely is a global necessity. Biodiversity provides the foundation for our agricultural systems. It provides the sources of traits to improve yield, quality, resistance to pests and diseases and adapt to changing environmental conditions, such as global warming. The protection of plant diversity is essential for food security and ecological well-being. Biodiversity is also a direct source of food for many people and is a essential part of our life support system. Without biodiversity our ecosystems, the planet's entire biosphere, cannot function.

Q13. What are approaches of conserving biodiversity?

Ans. There are different approaches to conserving biodiversity and different ways of using genetic resources. These are (i) on farm management, (ii) *In situ* conservation, (iii) *Ex situ* conservation and (iv) complementary conservation.

Q14. Describe on farm management of biodiversity.

Ans. On farm management involves the maintenance of crop species on the farm or in home gardens. The effectiveness of strategies to maintain and use crop or livestock diversity on farms depends on the extent to which local varieties continue to meet the needs of farmers and communities. Many plant genetic resources, especially those of minor crops, are managed as part of agricultural production systems. This type of biodiversity conservation has been termed 'conservation through use'. There are important reasons for supporting on farm maintenance of crop and livestock diversity:

   (i) It ensures the ongoing processes of evolution and adaptation of crops to their environments.

(ii) It allows for the continued selection of superior material by farmers that meets their needs and preferences.

(iii) It helps preserve indigenous knowledge, strengthens local institutions and promotes farmers' participation in national biodiversity conservation programmes.

(iv) It provides a necessary backup to gene bank collection.

(v) It provides natural laboratories for agricultural research.

The biodiversity of only those species can be maintained on the cultivators' farms and in home gardens which greater to the farmers and rural communities.

Q15. What is *In situ* conservation?

Ans. *In situ* (=on-site) conservation and use refers to the maintenance and use of wild plant populations in the habitats where they naturally occur and have evolved without the help of human beings. In other words, it refers to conservation of biodiversity under natural habitats. In this method, the wild species and the complete ecosystems are preserved together.

The wild populations regenerate naturally, and are dispersed naturally by wild animals, winds and in water courses. There exists an intricate relationship, often interdependence, between the different species and other components of the environment (such their pests and diseases) in which they occur. The evolution is purely driven by environmental pressures and any changes in one component affect the other. Provided that changes are not too drastic, this dynamic co-evolution leads to greater diversity and better adapted germplasm.

The main focus of Biodiversity's work is on the maintenance of the genetic diversity of the wild species and wild relatives of crop plants in the protected areas [national parks and nature reserves].

Q16. What are advantages and disadvantages of *in situ* conservation of biodiversity?

Ans. The advantages and disadvantages of *in situ* **conservation of biodiversity** are presented below:

Advantages

(i) The biodiversity is conserved under natural habitats.

(ii) It improves of the populations

(iii) It conserve complete ecosystem means several species together.

(iv) It protects endangered species in their original habitats.

Disadvantages

(i) It requires establishment of gene sanctuaries and national parks.

   (ii) Several areas have to be conserved to protect a single species.

  (iii) The management of protected ares poses several problems

  (iv) It is costly method of biodiversity conservation/

Q17. What is *Ex situ* conservation of biodiversity?

Ans. *Ex situ* (= off-site) conservation of germplasm takes place outside the natural habitat or outside the production system, in facilities specifically created for this purpose. Depending on the type of species to be conserved, different *ex situ* conservation methods may be used. The gene banks are used for *ex situ* conservation of biodiversity.

In gene banks, biodiversity is conserved for its sustainable use by breeders, farmers and researchers for agricultural development.

To make the genetic resources useful to farmers, breeders and researchers, gene bank managers must carefully document the collected materials, make the information available and establish a transparent and safe system for its distribution. They should take all the steps to make the material they conserve, including germplasm enhancement, is used by breeders and other researchers for agricultural development.

Q18. What are the advantages and disadvantages of *Ex situ* conservation of biodiversity?

Ans. The advantages and disadvantages of *Ex situ* conservation of biodiversity are given below:

Advantages

   (i) One species can be conserved at one place.

  (ii) The biodiversity is conserved either in seed gene banks or in field gene banks.

  (iii) Handling of material is easy.

  (iv) It requires less space for conservation.

Disadvantages

   (i) It requires establishment of seed gene banks and field gene banks.

  (ii) It require regular funding for maintenance of gene banks

  (iii) It involves high technical man power

  (iv) It requires periodical rejuvenation in case of seed crops

Q19. What are differences between *In situ* and *Ex situ* conservation of biodiversity?

Ans. There are several differences between *In situ* and *Ex situ* conservation of biodiversity which are presented in Table 13-2

TABLE 14-2: Comparison of *In situ* and *Ex situ* Conservation

Sl.No.	Particulars	In situ Conservation	Ex situ Conservation
1	Conservation place	Natural habitats	Gene banks
2	Basic requirements	Gene sanctuaries and national parks	Seed Gene banks and field gene banks
3	Skill required	Moderate	High
4	Species conserved	Several	Few at one place
5	Initial cost involved	Moderate	High
6	Management	Difficult	Easy
7.	Ecosystem conservation	Possible	Not possible

Q20. Describe briefly Complementary conservation of biodiversity.

Ans. It involves a proper combination of different conservation approaches for the sustainable use [in present and future] of genetic diversity existing in a target gene pool. The main objective of any plant genetic resources (PGR) conservation programme is to maintain the highest possible level of genetic variability present in the gene pool of a given species or crop both in its natural habitat and in a germplasm collection.

A proper combination of different methods depends on the species being conserved, the local infrastructure and human resources, the number of accessions in a given collection, its geographic site and intended use of the conserved germplasm. It does not advocate a particular method. A good complementary conservation strategy does not categorize crops or species into definitive classes. It is dynamic, and lends itself to meet the challenges of changes that are occurring in the field of genetic resources as it is open to new technologies and new needs.

Q21. What is Convention on Biological Diversity [CBD]?

Ans. The Convention on Biological Diversity is also known as the Biodiversity Convention. It is an international treaty that was adopted in Rio de Janeiro in June 1992. The Convention was opened for signature at the Earth Summit in Rio de Janeiro on 5 June 1992 and entered into force on 29 December 1993. The biodiversity agreement has been signed by 189 countries.

Q22. What are the main goals of CBD?

Ans. The Convention has following three main goals:

(i) Conservation of biological diversity (or biodiversity):

(ii) Sustainable use of its components; and

(iii) Fair and equitable sharing of benefits arising from genetic resources.

Q23. What is Cartagena Protocol?

Ans. The Cartagena Protocol on Biosafety of the Convention, also known as the Biosafety Protocol, was adopted in January 2000. The Biosafety Protocol seeks to protect biological diversity from the potential risks posed by living modified organisms resulting from modern biotechnology.

The Biosafety Protocol makes clear that products from new technologies must be based on the precautionary principle and allow developing nations to balance public health against economic benefits. It will for example let countries ban imports of a genetically modified organism if they feel there is not enough scientific evidence the product is safe and requires exporters to label shipments containing genetically altered commodities such as corn or cotton.

Q24. What are the main functions of the Conference of the Parties (COP)?

Ans. The convention's governing body is the **Conference of the Parties** (COP), consisting of all governments (and regional economic integration organizations) that have ratified the treaty. The main functiond of COP are to:

(i) Review progress under the Convention,

(ii) Identifies new priorities,

(iii) Sets work plans for members,

(iv) Make amendments to the Convention,

(v) Create expert advisory bodies,

(vi) Review progress reports by member nations, and

(vii) Collaborate with other international organizations and agreements.

The Conference of the Parties uses expertise and support from several other bodies that are established by the Convention. In addition to committees or mechanisms established on an ad hoc basis, two main organs are Secretariat and SBSTTA which are discussed below.

Q25. What are the function of the Secretariat of CBD?

Ans. The CBD Secretariat is based in Montreal [Canada]. It operates under the United Nations Environment Programme. Its main functions of the secretariat are to:

(i) Organize meetings,

(ii) Draft documents,

(iii) Assist member governments in the implementation of the programme of work,

(iv) Coordinate with other international organizations, and

(v) Collect and disseminate information.

Q26. What is the role SBSTTA?

Ans. The Subsidiary Body on Scientific, Technical and Technological Advice (SBSTTA). The SBSTTA is a committee composed of experts from member governments competent in relevant fields. It plays a key role in making recommendations to the COP on scientific and technical issues. Thirteenth Meeting of the Subsidiary Body on Scientific, Technical and Technological Advice (SBSTTA-13) held from 18 to 22 February 2008 in the FAO at Rome, Italy. SBSTTA-13 delegates met in the Committee of the Whole in the morning to finalize and adopt recommendations on the in-depth reviews of the work programmes on agricultural and forest biodiversity and SBSTTA's modus operandi for the consideration of new and emerging issues.

Q27. What is Indian Biodiversity Legislation?

Ans. In India the Biological Diversity Act [also called biodiversity act] was passed by the Central Government in 2002. The purpose of The Biodiversity Act - 2002 is to protect India's rich biodiversity [genetic resources] and associated knowledge against their use by foreign individuals, institutions or companies without sharing the benefits arising out of such use, and check bio-piracy.

Q28. Describe the Structures of Indian Biodiversity Act – 2002?

Ans. The biodiversity Act provides for establishment of a three tiered structure at the national, state and local level. It includes (i) National Biodiversity Authority (NBA), (ii) State Biodiversity Boards (SBB), and (iii) Biodiversity Management Committees (BMCs).

Q29. What is the role of National Biodiversity Authority (NBA)?

Ans. It deals with all matters relating to requests for access by foreign individuals, institutions or companies, and all matters relating to transfer of results of research to any foreigner.

Q30. What are the functions of State Biodiversity Boards (SBB)?

Ans. All matters relating to access by Indians for commercial purposes will be under the purview of the State Biodiversity Boards (SBB). The Indian industry will be required to provide prior intimation to the concerned SBB about the use of biological resource. The State Board will have the power to restrict any such activity, which violates the objectives of conservation, sustainable use and equitable sharing of benefits.

Q31. What is the role of Biodiversity Management Committees (BMCs)?

Ans. Institutions of local self government will be required to set up Biodiversity Management Committees in their respective areas for conservation, sustainable use, documentation of biodiversity and chronicling of knowledge relating to biodiversity. NBA and SBBs are required to consult the concerned BMCs on matters related to use of biological resources and associated knowledge within their jurisdiction.

Q32. Is there need for seeking permission of the National Biodiversity Authority for carrying out research?

Ans. There is no need for seeking permission of the National Biodiversity Authority for carrying out research, if it is carried out in India by Indians, as well as under collaborative research projects that have been drawn within the overall policy guidelines formulated by the Central Government. The only situations that would require permission of the NBA are: (i) when the results of any research which has made use of the country's biodiversity is sought to be commercialized, (ii) when the results of research are shared with a foreigner or foreign institution, and (iii) when a foreign institution/individual wants access to the country's biodiversity for undertaking research

Q33. What are the Exemptions of Indian Biodiversity Act 2002?

Ans. The legislation provides for the following exemptions:

(i) Exemption to local people and community of the area for free access to use biological resources within India

(ii) Exemptions to growers and cultivators of biodiversity and to Vaids and Hakims to use biological resources.

(iii) Exemption through notification of normally traded commodities from the purview of the Act

(iv) Exemption for collaborative research through government sponsored or government approved institutions subject to overall policy guidelines and approval of the Central Government

(v) The value added products have been excluded from the definition of biological resources.

(vi) The Indian researchers neither require prior approval nor need to give prior intimation to SBB for obtaining biological resource for conducting research in India.

Q34. What is the impact of Biodiversity Act on Indian industry?

Ans. The Indian industry is required to give prior intimation to the concerned SBB about obtaining the biological resources for commercial purposes. The SBB will have the power to prohibit or restrict any such activity, which violates the objectives of conservation, sustainable use and equitable sharing of benefits.

Q35. What is the impact of Biodiversity Act on Indian Ayurvedic industry?

Ans. The Act does not aim at banning the use of medicinal plants. It provides that for commercial use of resources and related knowledge by Indians only, prior intimation to the State Biodiversity Board is required. Hakims and vaids will continue to have free access to resources and knowledge.

Q36. Is there overlapping of Biodiversity Bill and Plant Varieties Protection Act ?

Ans. There is no overlap between Biodiversity Bill and Plant Varieties Protection (PVP) Act The scope and objectives of these two legislations are different The PVP legislation accords intellectual property rights to a person for developing a new plant variety. On the other hand, the biodiversity legislation is primarily aimed at regulating access to biological resources and associated knowledge so as to ensure equitable sharing of benefits arising from their use.

The Act provides that prior approval of NBA is necessary before applying for any kind of IPRs based on any research or information on a biological resource obtained from India. However, in case of patents, permission of the NBA may be obtained after acceptance of the patent but before sealing of the patent. This has been done to protect the priority date of the patent applicant.

Q37. What are Heritage sites?

Ans. The Act provides for designating heritage sites. These are areas of biodiversity importance, which harbour rich biodiversity, wild relatives of crops, or areas, which lie outside the protected area network. The purpose is not to cover the already designated protected areas such as national parks and wildlife sanctuaries.

Q38. What is the provision for threatened species?

Ans. The Act [Section 38] provides for notifying threatened species and prohibits or regulates their collection. It also provides for taking appropriate steps to rehabilitate and preserve those species, thereby ensuring their conservation and management.

Q39. What is the Check on Bio-piracy by Indian biodiversity Act 2002?

Ans. To check bio-piracy, the proposed legislation provides that access to biological resources and associated knowledge is subject to terms and conditions, which secure equitable sharing of benefits. Further, it would be required to obtain the approval of the National Biodiversity Authority before seeking any IPR based on biological material and associated knowledge obtained from India.

Q40. Who are the Benefit Claimers of biological resources ?

Ans. The benefit claimers are conservers of biological resources, creators and holders of knowledge and information relating to the uses of biological resources. While granting approvals, NBA will impose conditions, which secure equitable share in the benefits arising out of the use of biological resources occurring in India or knowledge relating to them. These benefits could include monetary gains, grant of joint ownership of IPRs, transfer of technology, association of Indian Scientists in R and D, setting up of venture capital fund *etc.*

In cases where specific individuals, or group of individuals are identifiable, the monetary benefits will be paid directly to them. Otherwise, the amount will be deposited in the National Biodiversity Fund.

Biosafety Legislations

Chapter 14

Cartagena Protocol on Biosafety

Q1. What is Cartagena Protocol on Biosafety?

Ans. The Cartagena Protocol on Biosafety is the first international agreement to regulate the trans-boundary movements of genetically engineered (GE) organisms. The Biosafety Protocol is a subsidiary agreement to the UN Convention on Biological Diversity (CBD), which was signed by over 150 governments at the Rio Earth Summit in 1992. The Biosafety Protocol seeks to protect biological diversity from the potential risks posed by living modified organisms [LMOs] resulting from modern biotechnology.

Biosafety refers to the safe application of modern biotechnology with regard to human health, animal health and environment. In other words, biosafety is a term used to describe efforts to reduce and eliminate potential risk resulting from biotechnology and its products.

Q2. Give a brief history of Cartagena Protocol on Biosafety.

Ans. The Cartagena Protocol arises from the United Nations Conference on Environment and Development (also known as the Earth Summit) held in Rio de Janeiro, Brazil, in June 1992, in which Convention on biological diversity was signed. The Cartagena Protocol on Biosafety, also known as the Biosafety Protocol, is a supplement of the convention on biological diversity. The Biosafety Working Group met six times between 1996 and 1999. The extraordinary meeting of the conference of parties [COP], was originally scheduled to adopt the Protocol in February 1999 in Cartagena city of Colombia. However, the meeting broke down because of disagreements regarding appropriate consideration of economic interests. The extraordinary meeting was temporarily suspended and only resumed after three more negotiation rounds achieved compromise. The Protocol was finally adopted by conference of parties [COP] in Montreal, Canada on 29 January 2000 and entered into force on 11 September 2003.

**Q3.	What is the need for biosafety protocol?**

Ans.	The need of such a strong Biosafety Protocol is illustrated by the genetic contamination of maize in Mexico. Genetically engineered cultivars of maize in Mexico resulted in contamination of non-transgenic varieties, land races and wild relatives of this major food crop. This was the first case of genetic pollution in a centre of origin and diversity of a major food crop. This invited the attention of crop scientists for adoption of strong biosafety protocol for handling of LMOs.

**Q4.	How many countries are members of Cartagena Protocol on Biosafety?**

Ans.	By June 2001, the Protocol had received 103 signatures. The membership is open to all countries of the World. The membership has been signed by 162 countries till November 2011. India signed the Cartagena Protocol in January 2001. However, countries like US, Argentina and Canada that produce about 90 percent of genetically engineered [GE] crops in the world have not ratified the Protocol, and are actively working to undermine it.

The required number of 50 instruments of ratification/accession/approval/acceptance by countries was reached in May 2003. As a result, the Protocol entered into force on 11 September 2003.

**Q5.	What are the functions of governing body of Cartagena Protocol on Biosafety?**

Ans.	The governing body of the Protocol is the Conference of the Parties to the Convention serving as the meeting of the Parties to the Protocol (COP-MOP). The main function of this body is to review the implementation of the Protocol and make decisions necessary to promote its effective operation. As of date, four meetings of the COP-MOP have been convened. The governing body includes Chairman and members from member countries. The Chairman is of the rank of Ambassador or High Commissioner of the member country. The term of the governing body is usually five years. The secretariat of the protocol is the same as for convention on biological diversity [CBD]. It is located at Montreal in Canada. Funds for the protocol are voluntarily contributed by contracting parties, developed country parties and other countries and sources. The main functions of governing body are:

(1)	To review the implementation of the Protocol and

(2)	To make decisions necessary to promote its effective operation.

**Q6.	What are the functions of Biosafety Protocol?**

Ans.	Biosafety protocol refers to various measures that are taken for the safe application of modern biotechnology. These are regulatory measures and policies that are designed to safeguard the public interest. The biosafety protocol became effective from January 2000. The main functions of biosafety protocol are as follows:

(1) **Promotion of Biosafety:** The Protocol promotes biosafety measures by establishing rules and procedures for the safe transfer, handling, and use of LMOs, developed through modern biotechnology.

(2) **Conservation of Biodiversity:** The Protocol applies to the trans-boundary movement [transit, handling and use] of all living modified organisms that may have adverse effects on the conservation and sustainable use of biological diversity. It ensures that LMOs should not have adverse effects on human health, animal health and biological diversity.

(3) **Regulation of World Trade:** Protocol is designed to regulate the international trade of genetically engineered organisms that may have adverse effects on the conservation and sustainable use of biological diversity, including risk to human health.

Q7. What are objectives and elements of Biosafety Protocol?

Ans. The objective of the Protocol is to ensure an adequate level of protection in the field of the safe transfer, handling and use of 'living modified organisms resulting from modern biotechnology' that may have adverse effects on the conservation and sustainable use of biological diversity, including risks to human health, specifically from trans-boundary movement of LMOs. The main elements of Cartagena biosafety protocol are briefly presented here as under.

(1) **Clearing House:** The Protocol establishes an Internet-based "Biosafety Clearing-House" to help countries exchange scientific, technical, environmental, and legal information about living modified organisms (LMOs).

(2) **Advance Informed Agreement Procedure:** It creates an advance informed agreement (AIA) procedure that in effect requires exporters to seek consent from an importing country before the first shipment of an LMO meant to be introduced into the environment (such as seeds for planting, fish for release, or microorganisms for bioremediation).

The AIA procedure does not apply to LMO commodities that are intended for food, feed, or processing (*e.g.*, maize, soy or cottonseed), to LMOs in transit, or to LMOs destined for contained use (*e.g.*, organisms intended only for scientific research within a laboratory).

(3) **Items not Addressed:** The Protocol does not address food safety issues. It does not pertain to non-living products derived from genetically engineered plants or animals, such as milled maize or other processed food products.

(4) **Documentation:** It requires shipments of LMO commodities, such as maize or soybeans that are intended for direct use as food, feed, or for processing, to be accompanied by documentation stating that such shipments "may contain" living modified organisms and are "not intended for intentional introduction into the environment. It also sets out information to be

included on documentation accompanying LMOs destined for contained use, including any handling requirements and contact points for further information and name and address of the importer and exporter. It should also include a declaration that the movement is in conformity with the Protocol and, as appropriate.

(5) Existing Rights Remain Unchanged: Parties must implement rights and obligations under the Protocol consistent with their existing international rights [such as WTO agreements] and obligations, including with respect to non-Parties to the Protocol.

(6) Capacity Building: The Protocol calls on Parties to cooperate with developing countries in building their capacity for managing modern biotechnology.

(7) Trade with non-Parties: The Protocol states that the "trans-boundary movement of living modified organisms between Parties and non-Parties shall be consistent with the objective of this Protocol.

Q8. What is Risk to Human and Animal Health from GMOs and their products?

Ans. The risk from genetically modified organisms and their products to human health include toxicity, allergenicity, antibiotic resistance and nutritional uptake. These are discussed as follows:

(i) Toxicity: Sometimes, the transgene leads to production of toxic substance by altering the metabolic pathway. Consumption of such transgenic products may cause toxic effect on human health. Hence, every transgenic product has to be evaluated for toxicity to human and animal health.

(ii) Allergenicity: Sometimes, transgenic products have allergic effects on human health. For example, transgenic soybean containing methionine gene from Brazil nut has allergic effects in sensitive persons. Hence, such soybean has not been approved for sale. Feeding of Bt. cotton seed to animals has not been reported to have any adverse effect. The Bt. cake also does not have any adverse effect on the digestion of animals. Moreover, no allergic or toxic effect of Bt. cotton seed and meal has been reported so far. The oil extracted from the seed of Bt. cotton has not been found to have any adverse effect on the human health.

(iii) Antibiotic Resistance: Antibiotic such as Kanamycin is used for selection of transgenic cells. It is believed that transgenic products when consumed will lead to resistance to such antibiotic. However, there is no such report so far.

(iv) Nutrient Uptake: There are possibilities that transgene product may inhibit absorption of some nutrients in human digestive system, resulting in adverse effects on human health. However, such evidences are not available.

Q9. **What is Risk to Environment from GMOs and their products?**

Ans. The risk from genetically modified organisms and their products to environment include development of super weeds, loss of biodiversity, contamination of non-transgenic cultivars, effect on non-target organisms, development of resistance, effect on soil ecology and evolution of new viruses. These aspects are briefly discussed as follows:

(i) **Development of Super Weeds:** Natural out crossing of transgenic plants with wild species and relatives of a crop may lead to development of super weeds. Such weeds may have herbicide resistance or insect resistance and will be more invasive than normal weeds. The out crossing is easy with other varieties or cross compatible species.

(ii) **Loss of Biodiversity:** Some farmers still maintain landraces and some old varieties which are valuable genetic resources. The natural ecosystem is still rich in plant biodiversity which needs to be conserved. The wide spread use of transgenic cultivars of different crops will lead to substantial loss of crop biodiversity.

(iii) **Contamination:** The cultivation of transgenic varieties will lead to contamination of landraces, non-transgenic cultivars and closely related wild species through natural out crossing. In Mexico, great diversity of both wild and cultivated corn is found. It has been reported that wild corn varieties in some remote areas of Mexico has been contaminated by Bt. transgenic cultivars.

The possibilities of cross pollination of Bt. cotton with other species are nil to negligible because the Bt. gene has been inserted in upland cotton ($2n = 52$) which cannot outcross with cultivated or wild diploid species ($2n = 26$). It can also not outcross with tetraploid wild species such as *G. tomentosum* which are found either in uncultivated areas or extremely isolated species garden maintained at different research institutes. The upland cotton in which Bt. gene has been inserted does not have cross compatibility with other genera of the family *Malvaceae*.

(iv) **Effect on Non-target Insects:** The genetically modified genes may sometimes have adverse effects on non-target insects/organisms. The adverse effects of Bt. insecticidal protein have been reported. No adverse effect of Bt. cotton has so far been reported on parasites, predators and other non-target beneficial insects such as honey bees, silk worm and Lac worm. However some adverse effects of Bt. gene have been reported on Monarch butterfly, Lice wings (an insect predator) and tobacco horn worm.

(v) **Development of Resistance:** There is a fear that over a period of time the target insect may develop resistance to Bt. insecticidal gene. However, in cotton there is no report on development of resistance to target insect so far. The first report of development of resistance to Bt. toxin is that of diamond back moth, an important pest of *Brassica* crops the world over.

(vi) Effect on Soil Ecology: It is considered that Bt. crops will liberate some additional chemicals from their roots and have adverse effects on soil ecology. Bt. [cry1 Ac] gene has been used in many crops, but no adverse effect of this gene on soil ecology has been reported so far.

(vii) Evolution of New Viruses: There is fear that development of virus resistant cultivars may sometimes lead to development of new stronger variants of virus that can affect transgenic plants.

Chapter 15

Biosafety Regulations

Q1. What is biosafety?

Ans. Biosafety refers to the safe application of modern biotechnology with regard to human health, animal health and environment. In other words, biosafety is a term used to describe efforts to reduce and eliminate potential risk resulting from biotechnology and its products.

Q2. What is biosafety protocol?

Ans. Biosafety protocol refers to various measures that are taken for the safe application of modern biotechnology. These are regulatory measures and policies that are designed to safeguard the public interest. The biosafety protocol became effective from January 2000. The main function of biosafety protocols are as follows:

(1) It ensures adequate level of protection for safe transfer, handling and use of living modified organisms[LMO] developed through modern biotechnology.

(2) It ensures that LMOs should not have adverse effects on human health, animal health and biological diversity.

(3) It looks into the trans boundary movements of living modified organisms.

Q3. What are the ways of risk assessment from genetically modified organisms?

Ans. The risk from genetically modified organisms and their products is assessed in two ways, *viz..* (i) risk to human and animal health, and (ii) risk to the environment.

Q4. What are risks to human and animal health from genetically modified organisms?

Ans. The risk from genetically modified organisms and their products to human health include toxicity, allergenicity, antibiotic resistance and nutritional uptake. These are discussed as follows:

(i) **Toxicity**: Sometimes, the transgene leads to production of toxic substance by altering the metabolic pathway. Consumption of such transgenic products cause toxic effect on human health. Hence, every transgenic product has to be evaluated for toxicity to human and animal health.

(ii) **Allergenicity**: Sometimes, transgenic products have allergic effects on human health. For example, transgenic soybean containing methionine gene from Brazil nut has allergic effects in sensitive persons. Hence, such soybean has not been approved for sale. Feeding of Bt. cotton seed to animals has not been reported to have any adverse effect. The Bt. cake also does not have any adverse effect on the digestion of animals. Moreover, no allergic or toxic effect of Bt. cotton seed and meal has been reported so far. The oil extracted from the seed of Bt. cotton has not been found to have any adverse effect on the human health.

(iii) **Antibiotic Resistance**: Antibiotic such as Kanamycin is used for selection of transgenic cells. It is believed that transgenic products when consumed will lead to resistance to such antibiotic. However, there is no such report so far.

(iv) **Nutrient Uptake**: There are possibilities that transgene product may inhibit absorption of some nutrients in human digestive system, resulting in adverse effects on human health. However, such evidences are not available.

Q5. What are risks to environment from genetically modified organisms?

Ans. The risk from genetically modified organisms and their products to environment include development of super weeds, loss of biodiversity, contamination of non-transgenic cultivars, effect on non-target organisms, development of resistance, effect on soil ecology and evolution of new viruses. These aspects are briefly discussed as follows:

(i) **Development of Super Weeds**: Natural out crossing of transgenic plants with wild species and relatives of a crop may lead to development of super weeds. Such weeds may have herbicide resistance or insect resistance and will be more invasive than normal weeds. The out crossing is easy with other varieties or cross compatible species.

(ii) **Loss of Biodiversity**: Some farmers still maintain landraces and some old varieties which are valuable genetic resources. The natural ecosystem is still rich in plant biodiversity which needs to be conserved. The wide spread use of transgenic cultivars of different crops will lead to substantial loss of crop biodiversity.

(iii) **Contamination**: The cultivation of transgenic varieties will lead to contamination of landraces, non-transgenic cultivars and closely related wild species through natural out crossing. In Mexico, great diversity of both wild and cultivated corn is found. It has been reported that wild corn varieties in some remote areas of Mexico has been contaminated by Bt. transgenic cultivars.

The possibilities of cross pollination of Bt. cotton with other species are nil to negligible because the Bt. gene has been inserted in upland cotton (2n = 52) which cannot outcross with cultivated or wild diploid species (2n = 26). It can also not outcross with tetraploid wild species such as *G. tomentosum* which are found either in uncultivated areas or extremely isolated species garden maintained at different research institutes. The upland cotton in which Bt. gene has been inserted does not have cross compatibility with other genera of the family *Malvaceae.*

(iv) **Effect on Non-target Insects**: The genetically modified genes may sometimes have adverse effects on non-target insects/organisms. The adverse effects of Bt. insecticidal protein have been reported. No adverse effect of Bt. cotton has so far been reported on parasites, predators and other non-target beneficial insects such as honey bees, silk worm and Lac worm. However some adverse effects of Bt. gene have been reported on Monarch butterfly, Lice wings (an insect predator) and tobacco horn worm.

(v) **Development of Resistance**: There is a fear that over a period of time the target insect may develop resistance to Bt. insecticidal gene. However, in cotton there is no report on development of resistance to target insect so far. The first report of development of resistance to Bt. toxin is that of diamond back moth, an important pest of *Brassica* crops the world over.

(vi) **Effect on Soil Ecology**: It is considered that Bt. crops will liberate some additional chemicals from their roots and have adverse effects on soil ecology. Bt. [cry1 Ac] gene has been used in many crops, but adverse effects of this gene on soil ecology has not been reported so far.

(vii) **Evolution of New Viruses**: There is fear that development of virus resistant cultivars may sometimes lead to development of new stronger variants of virus that can effect transgenic plants.

Q6. What are biosafety regulatory mechanisms in India?

Ans. India has a well-defined regulatory mechanism for development and evaluation of genetically modified organisms [GMOs] and their products. The department of biotechnology [DBT] and the Ministry of Environment and Forests [MoEF] are the two apex regulatory bodies. Guidelines for safety have been issued by the Department of Biotechnology [DBT] in 1990 covering research in biotechnology, field trials and commercial applications. Presently, there are following six competent authorities for implementation of regulations and guidelines in the country.

(1) Recombinant DNA Advisory Committee [RDAC],

(2) Review Committee on Genetic manipulation [RCGM],

(3) Genetic Engineering Approval Committee [GEAC]- an apex body,

(4) Institutional Biosafety Committee [IBSC] attached to every organization engaged in recombinant DNA research,

(5) State Biosafety Coordination Committees [SBCC}, and

(6) District Level Committees [DLC].

Q7. What is the role of recombinant DNA advisory committee [RDAC]?

Ans. This committee constituted by the Department of Biotechnology takes note of development in biotechnology at national and International levels. RDAC prepares recommendations from time to time that are suitable for implementation for upholding the safety regulations in research and applications of GMOs and their products. This committee prepared the Recombinant DNA Biosafety Guidelines in 1990, which was adopted by the Government of India for conducting research and handling of GMOs in India.

Q8. What is the role of review committee on genetic manipulation [RCGM]?

Ans. The RCGM under the Department of Biotechnology has the following functions.

(i) To bring out manuals of guidelines specifying producers for regulatory process on GMOs in research, use ans applications including industry with a view to ensure environmental safety.

(ii) To review all ongoing r-DNA projects involving high risk, category and controlled field experiments.

(iii) To lay down procedure for restriction or prohibition, production, sale, import and use of GMOs both for research and applications.

(iv) To permit experiments with category II risks and above with appropriate containment.

(v) To authorize imports of GMOs/transgenes for research purposes.

(vi) To authorize field experiments in 20 acres in multi-location in one crop season with up to one acre at one site.

(vii) To generate relevant data on transgenic material in appropriate systems.

(viii) To undertake visits of sites of experimental facilities periodically, where projects with biohazard potentials are being pursued and also at a time prior to the commencement of the activity to ensure that adequate safety measures are taken as per the guidelines.

Q9. What is the role of genetic engineering approval committee [GEAC]?

Ans. It functions as a body under Ministry of Environment and Forests and is responsible for approval of activities involving large scale use of hazardous micro-organisms and recombinant products in research and industrial production from the environment angle. Its main functions are as follows:

(i) To permit the use of GMOs and their products for commercial application.

(ii) To adopt producers for production, sale, import and use of GMOs both for research and applications under Environment Protection Authority [EPA].

(iii) To authorize large scale production and release of GMOs and their products into the environment.

(iv) To authorize agencies or persons to have powers to take punitive action under the EPA.

Q10. What is the role of institutional biosafety committee [IBSC]?

Ans. The IBSC is the nodal point for interaction with institution for implementation of the guidelines. It is constituted before undertaking any project involving manipulation of micro-organisms, plants or animals. The main activities of IBSC are as follows:

(i) To note and approve r-DNA work.

(ii) To ensure adherence of r-DNA safety guidelines of government.

(iii) To prepare emergency plan according to guidelines.

(iv) To recommend to RCGM about category II risk or above experiments and to seek approval of RCGM.

(v) To act as a nodal point for interaction with statutory bodies.

(vi) To ensure experimentation at designated locations, taking into account approved protocol.

Q11. What is the role of State Biosafety Coordination Committees[SBCC]?

Ans. This committee is constituted in each state where research and application of GMO is carried out. It is headed by Chief Secretary of the State. Its main functions are as follows:

(i) To ensure adherence to r-DNA safety guidelines issued by the government.

(ii) To review periodically the safety and control measures in various institutions handling GMOs.

(iii) To act as nodal agency at State level to assess the damage, if any, due to release of GMOs and to take on site control measure.

(iv) To coordinate activities related to GMOs in the state with central Ministries.

(v) To nominate State Government representative in the activities related to field inspection of GMOs.

Q11. What is the role of District Level Committees [DLC]?

Ans. This is constituted at the district level and is the smallest committee. It is headed by the District Collector. Its functions are as follows:

(i) To monitor the safety regulations in the institutions.

(ii) It has powers to inspect, investigate and report to the SBCC or GEAC about compliance or non compliance of r-DNA guidelines or violations under EPA.

(iii) To act as nodal agency at District level to assess the damage, if any, due to release of GMOs and to take on site control measures.

Miscellaneous Topics

Chapter 16

Plant Quarantine Order, 2003

Q1. What is Plant Quarantine Order, 2003?

Ans. The Plant Quarantine order, 2003 was approved by the government of India, Ministry of Agriculture (Department of Agriculture and Cooperation) on 18th[h] November, 2003. This order may be called the Plant Quarantine (Regulation of Import into India) Order, 2003. This order has replaced the Plants, Fruits and Seeds order, 1989. This Order has been enacted for the purpose of prohibiting and regulating the import into India of agricultural articles. This order consists of seven chapters. Chapter 1 deals with title, scope and definitions of various terms. Chapter 2 deals with general conditions of import of seed, fruits and plants. Chapter 3 explains special conditions of import. Chapter 4 deals with post entry quarantine, Chapter 5 deals with appeals and revision. Chapter 6 explains powers of relaxation and Chapter 7 deals with repeal and savings.

This order may be called the Plant Quarantine (Regulation of Import into India) Order, 2003. This Order comes into force with effect from 1st January, 2004. It extends to whole of India

Q2. What are general conditions for import as per Plant Quarantine Order, 2003?

Ans. General conditions for import as per Plant Quarantine Order, 2003 are as follows:

(1) **Valid Permit**: As per this order, no consignment of plants and plant products and other regulated articles shall be imported into India without a valid permit issued by competent authority.

(2) **Material for Import:** Plant materials of specified categories and species can be imported from specified countries only. Commercial import of consignments of seeds of coarse cereals, pulses, oil seeds and fodder seeds and seeds/stock material of fruit plant species for propagation shall

only be permitted based on the recommendations of EXIM Committee of Department of Agriculture and Cooperation, except the specified trial material.

(3) **Application:** Every application for a permit shall be made at least one month in advance to the Issuing Authority, in Form PQ 01 for the import of plants and plant products for consumption and processing and in form PQ 02 for import of seeds and plants for propagation.

(4) **Fees:** A fee of Rs.150/- shall be payable along with the application for the import of seeds, fruits and plants for consumption and Rs.300/- for application for the import of seeds and plants for sowing or planting and the fee shall be payable in the form of Demand Draft payable to the Competent Authority having jurisdiction.

(5) **Issue of Permit:** The Issuing Authority shall issue permit in quadruplicate in form PQ 03 for import of plants and plant products for consumption and in form PQ 04 for import of seeds and plants for sowing or planting, if he is satisfied that the applicant meets all the necessary conditions. One copy of import permit shall be forwarded to the exporter in advance to facilitate incorporation of import permit number in the phytosanitary certificate issued by the exporting country. The import permit shall be issued for import of specified material from specified countries only.

(6) **Period of Validity:** The import permit issued shall be valid for a period of six months from the data of issue and valid for successive shipment provided the exporter and importer, bill of entry, country of origin and phytosanitary certificate are the same for the entire consignment. The issuing authority may, on request, extend the period of validity for a further period of six months after charging Rs. 200/- and Rs. 100/- as revalidation fee for propagation and consumption plant material respectively provided such request for extension of validity is made to the issuing authority before the expiry of the permit with adequate reasons to be recorded in writing.

(7) **Quantity for Import:** The quantity mentioned in the import permit if exceeds by up to 10 per cent may be allowed by charging additional inspection fee and import permit fee provided the excess quantity reflected in the phytosanitary certificate from the country of exporting. The import permit will become invalid if quantity exceeds more than 10 per cent of the quantity of import permit. Suppression of the facts or any material information while issue of import permit is liable to be cancelled or withdrawn.

(8) **Transfer:** The import permit issued shall not be transferable and no amendments to the permit shall be issued except for change of point of entry subject to reasons to be recorded in writing.

(9) **Tag Color:** An orange and green color tag shall be issued in form PQ 05 in the case of permits issued for import of seeds and plants for sowing or

planting so as to facilitate the identification of consignments at the time of their arrival at the point of entry.

(10) Contamination: No consignment of seed or grain shall be permitted to be imported with contamination of quarantine weeds. Every application for quarantine inspection and clearance shall be made in Form PQ 15.

(11) Places of Import: All the consignments of plants and plant products and other regulated articles shall be imported into India only through ports of entry as specified in Schedule-I and Inland Container, Depots/Container Freight Stations and foreign post offices falling within the jurisdiction of concerned plant quarantine station operating here under or those notified by the Government from time to time in this behalf.

All consignments of seeds and plants for propagation and regulated articles such as live insects, microbial cultures, bio-control agents and soil shall only be imported into India through regional plant quarantine stations of Amritsar, Chennai, Kolkata, Mumbai or New Delhi or through any other points of entry as may be notified from time to time for this purpose. However, no import of germplasm/transgenic plant material and genetically modified organisms shall be permitted through New Delhi Airport.

(12) Inspection of material: On arrival, at the first point of entry the consignment shall be inspected by the Plant Protection Adviser or any other officer duly authorized by him in this behalf and appropriate samples shall be drawn for laboratory testing, in accordance with the guidelines issued by Plant Protection Adviser from time to time.

(13) Quarantine Clearance: The Plant Protection Adviser or the officer authorized by him may, after inspection and laboratory testing, fumigation, disinfection or disinfestation, as may be considered necessary by him, accord quarantine clearance for the entry of a consignment or grant provisional clearance for growing under post-entry quarantine, as the case may be in form PQ 16 and or order deportation or destruction of the consignment in form PQ 17 in the event of non-compliance with the restrictions and conditions specified in this Order.

(14) Cost of Fumigation, *etc.***:** Where fumigation or disinfestation or disinfection is considered necessary in respect of a consignment of plants, seeds and fruits the importer shall on his own and at his cost arrange for the fumigation, disinfection or disinfestation of the consignment, through an agency approved by the Plant Protection Adviser under the supervision of an officer duly authorized by the Plant Protection Adviser in that behalf.

(15) Responsibilities of Importer: The importer of seed material has various responsibilities. Importer himself or his authorized agent has following responsibilities,

    (i) To file an application for the quarantine inspection of imported seeds, plants and plant products or other regulated articles in the form PQ 15 along with copies of relevant documents and fees as prescribed under Schedule-IX payable by a demand draft to the competent authority

(ii) To provide information on any plant and plant product and other articles covered under this Order and which are imported by him/her or are in his/her possession, to Plant Protection Adviser or any officer duly authorized by him;

(iii) To bring the consignments to the concerned plant quarantine station or to place of inspection, fumigation or treatment as directed by Plant Protection Adviser or any officer duly authorized by him;.

(iv) To permit drawing of appropriate samples for inspection and laboratory investigation and extend necessary facilities towards the same;

(v) To open, repack and load into or unload from the fumigation chamber and seal the consignment;

(vi) To remove them after inspection and treatment according to the directions issued by the Plant Protection Adviser or any officer authorized by him;

(vii) To arrange deportation or destruction of the consignment at the cost of importer as may be deemed necessary by Plant Protection Adviser or an officer authorized by him

(16) Container: The container carrying plants and plant products for import from other countries should be packed properly. It should not permit spillage of material or contamination with soil or escape of any pest. Moreover, the package or container shall not be opened or seals are broken anywhere in India.

(17) Phytosanitary Certificate: No consignment shall be permitted import unless accompanied by a Phytosanitary Certificate issued by an authorized officer at the country of origin in the form PQ 21 or at the country of re-export in form PQ 22;

However, cut flowers, garlands, bouquets, dry fruits/nuts *etc.*, weighing not more than two kilograms imported for personal consumption may be allowed to be imported without a Phytosanitary Certificate or an import permit.

(18) Packaging: The packed material that is appropriately treated permitted for import. The treatments shall include heat-kiln treatment at 560 C for a minimum of 30 hrs or Methyl Bromide fumigation at 48 g/cum for 32 hours or chemical impregnation of wood with wood preservatives such as copper chrome arsenic or any other approved treatment as per international standards and the treatment shall be endorsed in phytosanitary certificate. No article packed with packaging materials shall be released by the proper officers of customs unless the consignment is accompanied by a phytosanitary certificate in respect of said packing material; If no phytosanitary certificate is furnished in respect of said packaging material, the proper officer of customs shall grant out of charge only after clearance is obtained from local plant quarantine authorities, who shall grant clearance from the quarantine angle and may, if deemed fit, subject the said packaging material to treatment at the expense of importer.

(19) Import of soil, *etc.* - No import of soil, earth, compost, sand, plant debris along with plants, fruits and seeds shall be permitted except under the following conditions:

(i) The consignments of soil, earth, clay and similar material for any microbiological, soil mechanics, or mineralogical investigations and peat for horticultural purposes may be permitted through specified air or sea ports or land custom station, on applications made for that purpose;

(ii) The application for the purpose referred to in (i) above shall be made to the Plant Protection Adviser, at least one month in advance, in form PQ 06 along with a registration fee of Rs. 200/- by a bank draft drawn in favor of Accounts Officer, Directorate of Plant Protection, Quarantine and Storage, Faridabad.

(iii) The Plant Protection Adviser may, after scrutiny of the application, and if satisfied of the purpose, for which such consignment is being imported, issue special permit in Form PQ

(iv) The consignments soil, peat or sphagnum moss *etc.*, shall be inspected, fumigated, disinfected or disinfested by the importer from an agency approved by the Plant Protection Adviser under the supervision of an officer duly authorized by Plant Protection Adviser.

(20) Fees for inspection, fumigation, *etc.* The importer of the consignment or his agent shall pay to the Plant Protection Adviser or any other officer duly authorized by him in this behalf, the fees prescribed for inspection, fumigation, disinfestation, disinfection of consignment.

(21) Permits required for import of Germplasm, Transgenic or Genetically Modified Organisms:

(i) No consignment of germplasm/transgenics/Genetically Modified Organisms (GMOs) shall be imported into India for research/ experimental purpose without valid permit issued by the Director, National Bureau of Plant Genetic Resources, New Delhi.

(ii) Every application for import of plant germplasm/transgenics/ genetically modified organisms for research/experimental purpose by the public/private organizations will be made to the Director, National Bureau of Plant Genetic Resources, New Delhi in form PQ 08 and the permit shall be issued in form PQ 09 in triplicate and a red/green tag in PQ 10 for germplasm and a Red/White tag in PQ 11 for transgenic/Genetically Modified Organisms. Such permits for import of transgenic/Genetically Modified Organisms shall be issued subject to the approval of Review Committee on Genetic Manipulation (RCGM) set- up by Department of Biotechnology.

(iii) No imported consignments of plant germplasm/transgenics/ genetically modified pests shall be opened at the point of entry and it shall be forwarded to the Director, National Bureau of Plant Genetic Resources, New Delhi.

(22) Permit required for import of live insects and microbial cultures:

(i) No Consignment of live insects, microbial cultures or bio-control agents shall be permitted into India without valid import permit issued by the Plant Protection Adviser.

(ii) Every application for permit to import insects or microbial cultures including algae or bio-control agents, shall be made in the form PQ12 at least two months in advance to Plant Protection Adviser along with a fee of Rs. 200/- towards registration in the form of bank draft issued in favor of the Accounts Officer, Directorate of Plant Protection Quarantine and Storage, Faridabad.

(iii) The Plant Protection Adviser shall issue the permit in Form PQ13 in triplicate, if satisfied of the purpose for which import is made and subject to such conditions imposed thereon. A yellow- green color tag or label in the form PQ14 shall be issued which shall be affixed on the parcel at the time of export.

(iv) All the consignments of insects, microbial cultures and bio-control agents shall be permitted only through specified points of entry. The consignment of beneficial insects shall be accompanied by a certificate issued by National Plant Protection Organization at the country of origin with additional declarations for freedom from specified parasites and parasitoids and the bio-control agents free from hyper-parasites. The consignment of beneficial insects/bio-control agents shall be subjected to post-entry quarantine as may be prescribed by the Plant Protection Adviser.

(23) Permit required for import of plants and plant products

(i) No consignment of plants and plant products, if found infested or infected with a quarantine pest or contaminated with noxious weed species shall be permitted to be imported.

(ii) Every vessel carrying out bulk shipment of grains shall be inspected on board by an officer duly authorized by Plant Protection Adviser before the same accorded permission to offload the grain at the notified port of entry. On inspection, if found free from quarantine pests and noxious weed species, permission shall be accorded to off- load the grain at the port or order fumigation/treatment of grain on board or immediately upon unloading at the port, as the case may be, before such permission is granted for movement outside the port and subject to such conditions as imposed thereon.

(iii) The bulk shipment (s) of transgenic plants or plant products or genetically modified organisms shall be dealt as per the provisions of the Rules for manufacture, use, import, export and storage of hazardous micro-organisms,

(24) Permit for import of timber

(i) No consignment of timber shall be permitted import unless the following conditions and requirements are fulfilled.

(a) The timber shall be stripped off its bark, either be squared or rounded and accompanied by an official statement that the wood has been appropriately fumigated/treated and such treatment shall be endorsed in the phytosanitary certificate issued at the country of origin or re-export, as the case may be;

(b) The timber shall be marked 'kiln-dried' or with any internationally recognized mark.

(ii) All the consignments of timber shall be inspected on board prior to unloading at the port of arrival by an officer duly authorized by PPA and, if necessary, fumigated/treated on board before unloading.

However, the above conditions shall not apply for containerized cargo, which shall be inspected by an authorized PQ officer after unloading of the containers from the ship at the port/container freight station or Inland Container Depots under the jurisdiction of concerned Plant Quarantine Station.

Q3. What are special conditions for import as per Plant Quarantine Order, 2003?

Ans. The general conditions will apply to import of all plant materials. In addition to the general conditions listed above, there are some specific conditions for import permit issued by an appropriate issuing authority.

Every consignment of plant species shall be accompanied by a Phytosanitary Certificate. The certificate is issued by the authorized officer at country of origin or by the country of re-export along with attested copy of original phytosanitary certificate. It has to declare that material is free from specified pests or that the specified pests do not occur in the country or state of origin. It should be supported by documentary evidence.

Q4. What are guidelines for Post entry Quarantine as per Order, 2003?

Ans. Plants and seeds, which require post-entry quarantine will be grown in post-entry quarantine facilities duly established by importer at his cost, approved and certified by the Inspection Authority as per the guidelines prescribed by the Plant Protection Adviser. The period for which, and the conditions under which, the plants and seeds shall be grown in such facilities shall be specified in the permit.

(i) PET shall not apply to the import of tissue-cultured plants that are certified virus-free, but such plants, shall be subjected to inspection at the point of entry to ensure that the phytosanitary requirements are met with.

(ii) Every application for certification of post-entry quarantine facilities shall be submitted to the inspection authority in Form PQ 18. The inspection authority if satisfied after necessary inspection and verification of facilities shall issue a certificate in Form PQ 19.

(iii) At the time of arrival of the consignment, the importer shall produce this certificate before the Officer-in-Charge of the Quarantine Station at the entry point along with an undertaking in form PQ 20.

(iv) If the Officer- in-Charge of the Quarantine Station, after inspection of the consignment is satisfied, he shall accord quarantine clearance with post-entry quarantine condition on the production, by an importer, of a certificate from the inspection authority with the stipulation that the plants shall be grown in such post-entry quarantine facility for the period specified in the import permit.

(v) After according quarantine clearance with post-entry quarantine conditions to the consignments of plants and seeds requiring post-entry quarantine, the Officer-in-Charge of the Quarantine Station at the entry point shall inform the inspection authority, having jurisdiction over the post-entry quarantine facility, of their arrival at the location where such plants would be grown by the importer.

Q5. What are responsibilities of importer as per Quarantine Order, 2003?

Ans. The importer or his authorized agent shall be responsible for the following.

(i) To intimate the inspection authority in advance about the date of planting of the imported plant or seed.

(ii) Not to transfer or part with or dispose the consignment during the pendency of post entry quarantine except in accordance with a written approval of inspection authority.

(iii) To permit the inspection authority complete access to the post-entry quarantine facility at all times and abide by the instructions of such inspection authority.

(iv) To maintain an inspection kit containing all requisite items to facilitate nursery inspection and ensure proper plant protection and upkeep of nursery records.

(v) To extend necessary facilities to the inspection authority during his visit to the nursery and arrange destruction of any part or whole of plant population when ordered by him in the event of infection or infestation by a quarantine pest, in a manner specified by him.

(vi) The inspection authority of concerned area of jurisdiction or any officer authorized by the Plant Protection Adviser in this behalf, in association with a team of experts shall inspect the plants grown in the approved post-entry quarantine facility at such intervals as may be considered necessary in accordance with the guidelines issued by the Plant Protection Adviser, with a view to detect any pests and advise necessary phytosanitary measures to contain the pests.

(vii) The inspection authority shall permit the release of plants from post-entry quarantine, if they are found to be free from pests and diseases for the period specified in the permit for importation.

(viii) Where the plants in the post-entry quarantine are found to be affected by pests and diseases during the specified period the inspection authority shall:

    (a) Order the destruction of the affected consignment of whole or a part of the plant population in the post-entry quarantine if the pest or disease is exotic, or

    (b) Advise the importer about the curative measures to be taken to the extent necessary, if the pest or disease is not exotic and permit the release of the affected population from the post-entry quarantine only after curative measures have been observed to be successful. Otherwise, the plants shall be ordered to be destroyed.

(ix) Where destruction of any plant population is ordered by the inspection authority, the importer shall destroy the same in the manner as may be directed by the inspection authority and under his supervision

(x) At the end of final inspection, the inspection authority shall forward a copy of the report of post-entry quarantine inspection duly signed by him to the Plant Protection Adviser under intimation to officer- in-charge of concerned plant quarantine station.

(xi) The importer shall be liable to pay the prescribed fee for inspection of plants in the Post-entry Quarantine facility

Q6. Explain procedure of appeal and revision as per Quarantine Order, 2003.

Ans. **(i) Appeal:** If an importer is aggrieved by the decision of the inspection authority regarding the destruction of any plant population, he may appeal to the Plant Protection Adviser within 7 days from the date of communication of the decision giving the grounds of appeal.

    (a) It shall be lawful for the Plant Protection Adviser to rely on the observations of the inspection authority and such other expert opinion, as he may deem necessary, for deciding the appeal.

    (b) The memorandum of appeal shall be accompanied by a bank draft in favor of the Plant Protection Adviser and payable at Faridabad, evidencing the payment of fee of Rs. 100/-

(ii) Revision: The Plant Protection Adviser may, at any time, call for the records relating to any case pending before the inspection authority for the purpose of satisfying itself as to the legality or propriety of any decision passed by that authority and may pass such order in relation thereto, as it thinks fit: But no such order shall be passed after the expiry of three months from the date of the decision; Moreover, the Plant Protection Adviser shall not pass any order prejudicial to any person, without giving him a reasonable opportunity of being heard.

Q7. What are powers of relaxation as per Quarantine Order, 2003?

Ans. There are certain conditions under which Relaxation of Import Permit and Phytosanitary Certificate is allowed as given below:

(i) The Central Government may, in public interest, relax any of the conditions of this Order relating to the import permit and the phytosanitary certificate in relation to the import of any consignment. The Joint Secretary in-charge of Plant Protection in the Department of Agriculture and Cooperation shall be the competent authority for according the relaxation.

(ii) In the event of grant of relaxation by competent authority, the consignment shall be released after charging the fee for import permit and fee for plant quarantine inspection at five times of normal rates.

(iii) The provisions of this Order shall apply without prejudice to the Customs Act, 1962 or any other Acts or Order related to imports.

Q8. Give a comparison of seed order 1989 and plant quarantine order, 2003

Ans. There are some similarities and some dis-similarities between seed order, 1989 and plan quarantine order, 2003. A brief comparison of these two orders is presented in the Table 16.1.

TABLE 16-1: Comparison of Seed Order, 1989 and Plant Quarantine Order, 2003

Sl.No.	Particulars	Seed Order, 1989	Plant Quarantine Order, 2003
1	Permit	Compulsory	Compulsory
2	Permit Fee for plants, fruits and seed used for consumption	Rs.50/	Rs. 150/
3	Permit Fee for plants, fruits and seed used for planting	Rs. 100/	Rs. 300/
4	Validity of permit	Six months	Six months
5	Extension of permit	Up to six months	Six months
6	Fees for extension of permit	Rs. 50/and Rs. 100/	Rs. 100/and Rs. 200/
7	Form used for Issue of permit for consumption	Form C	PQ 03
8	Form used for Issue of permit for planting	Form D	PQ 04
9			
10	Import of material through	Specified sea and airports	Specified sea and airports

Sl.No.	Particulars	Seed Order, 1989	Plant Quarantine Order, 2003
11	Post entry quarantine	Applicable	Applicable
12	Fee for post entry quarantine inspection to be paid by	Importer	Importer
13	Appeal and revision	Applicable	Applicable
14	Power of relaxation	Vested with government	Vested with government

Q9. What are the orders and notifications which can be repealed?

Ans. The following orders and notifications are hereby repealed,

(i) Rules for regulating the import of insects into India.

(ii) Rules for regulating the import of fungi into India.

(iii) Import of cotton into India

(iv) Plants, Fruits and Seeds Order, 1989

An import permit issued by any competent authority, which is in force immediately before the commencement of this Order and which is consistent with this Order, shall continue in force and all appointments made and fees levied under the repealed Rules, Regulations and Orders, and in force immediately before such commencement shall likewise continue in force and be deemed to be made or levied in pursuance of this Order until revoked.

Appendices

Appendix 1. List of Agricultural Universities in India

Sl.No.	Name of University	Location	State
	I. AGRICULTURAL UNIVERSITIES		
A	**STATE UNIVERSITIES**		
1	Acharya NG Ranga Agricultural University	Hyderabad	Telangana
2	Assam Agricultural University	Jorhat	Assam
3	Bihar Agricultural University	Sabour	Bihar
4	Birsa Agricultural University	Ranchi	Jharkhand
5	Indira Gandhi Krishi Vishwavidyalaya	Raipur	Chhatishgarh
6	Anand Agricultural University	Anand	Gujarat
7	Junagadh Agricultural University	Junagarh	Gujarat
8	Navsari Agricultural University	Navsari	Gujarat
9	Sardarkrushinagar-Dantiwada Agricultural University	Bansakantha	Gujarat
10	Chaudhary Charan Singh Haryana Agricultural University	Hisar	Haryana
11	CSK Himachal Pradesh Krishi Vishvavidyalaya,	Palampur	Himachal Pradesh
12	Sher-E-Kashmir Univ of Agricultural Sciences and Technology	Jammu	Jammu and Kashmir
13	Sher-E-Kashmir Univ of Agricultural Sciences and Technology of Kashmir	Srinagar	Jammu and Kashmir
14	University of Agricultural Sciences	Bangalore	Karnataka
15	University of Agricultural Sciences	Dharwad	Karnataka
16	University of Agricultural Sciences	Raichur	Karnataka
17	Kerala Agricultural University	Vallanikara	Kerala
18	Jawaharlal Nehru Krishi Viswavidyalaya	Jabalpur	Madhya Pradesh
19	Rajmata Vijayaraje Scindia Krishi Viswavidyalaya	Gwalior	Madhya Pradesh
20	Dr Balasaheb Sawant Konkan Krishi Vidyapeeth	Ratnagiri	Maharashtra
21	Dr Panjabrao Deshmukh Krishi Vidyapeeth	Akola	Maharashtra
22	Mahatma Phule Krishi Vidyapeeth	Rahuri	Maharashtra
23	Marathwada Agricultural University	Parbhani	Maharashtra

Sl.No.	Name of University	Location	State
24	Odisha Univ of Agriculture and Technology	Bhubaneswar	Odisha
25	Punjab Agricultural University	Ludhiana	Punjab
26	Maharana Pratap Univ of Agriculture and Technology	Udaipur	Rajasthan
27	Rajasthan Agricultural University	Bikaner	Rajasthan
28	Agriculture University	Kota	Rajasthan
29	Sri Karan Narendra Agriculture University	Jobner	Rajasthan
30	Agriculture University	Jodhpur	Rajasthan
31	Tamil Nadu Agricultural University	Coimbatore	Tamil Nadu
32	Chandra Shekar Azad University of Agriculture and Technology	Kanpur	Uttar Pradesh
33	Acharya Narendra Deva University of Agriculture and Technology	Ayodhaya	Uttar Pradesh
34	Sardar Ballabh Bhai Patel Univ of Agriculture and Technology	Meerut	Uttar Pradesh
35	Manyavar Kashiram University of Agriculture and Technology	Banda	Uttar Pradesh
36	Govind Ballabh Pant University of Agriculture and Technology	Pantnagar	Uttrakhand
37	Bidhan Chandra Krishi Viswavidyalaya	Kalyani	West Bengal
38	Uttar Banga Krishi Viswavidyalaya	Cooch Bihar	West Bengal
B	**CENTRAL UNIVERSITIES**		
1	Central Agricultural University	Imphal	Manipur
2	Rani Laxmibai Central Agricultural University	Jhansi	Uttar Pradesh
3	Rajendra Central Agricultural University	Pusa	Bihar
C	**DEEMED UNIVERSITIES**		
1	Indian Agricultural Research Institute	Pusa	New Delhi
2	National Dairy Research Institute	Karnal	Haryana
3	Indian Veterinary Research Institute	Bareilly	Uttar Pradesh
4	Central Institute for Fisheries Education	Mumbai	Maharashtra
5	Allahabad Agricultural Institute	Prayagraj	Uttar Pradesh
II. ANIMAL SCIENCE UNIVERSITIES			
S. N.	Name of University	Location	State
1	Sri Venkateswara Veterinary University	Tirupati	Andhra Pradesh

Sl.No.	Name of University	Location	State
2	Karnataka Veterinary, Animal and Fisheries Sciences University	Bidar	Karnataka
3	Maharashtra Animal Science and Fishery University	Nagpur	Maharashtra
4	Guru Angad Dev Veterinary and Animal Science University	Ludhiana	Punjab
5	Tamil Nadu Veterinary and Animal Science University	Chennai	Tamil Nadu
6	Tamil Nadu Fisheries University	Nagapattinam	Tamil Nadu
7	UP Pandit Deen Dayal Upadhaya Pashu Chikitsa Vigyan Vishwa Vidhyalaya evam Go Anusandhan Sansthan, Mathura	Mathura	Uttar Pradesh
8	West Bengal University of Animal and Fishery Sciences	Kolkata	West Bengal
9	Rajasthan University of Veterinary and Animal Sciences	Bikaner	Rajasthan
10	Kerala University of Fisheries and Ocean Studies	Kochi	Kerala
11	Kerala Veterinary and Animal Sciences University	Wayanand	Kerala
III. HORTICULTURAL UNIVERSITIES			
1	Andhra Pradesh Horticultural University	Tadepalli-gudem	Andhra Pradesh
2	University of Horticultural Sciences	Bagalkot	Karnataka
3	Dr Yashwant Singh Parmar Univ of Horticulture and Forestry	Solan	Himachal Pradesh
4	Dr YSR Horticultural University	Veketara-manna-gudem	Andhra Pradesh
5	Uttarakhand University of Horticulture and Forestry	Bharsar	Uttrakhand

Appendix 2. List of ICAR Research Institutes in India

Sl.No.	Name of Research Institute	Location	State
A	**CROP SCIENCES**		
1	ICAR-Central Institute for Cotton Research	Nagpur	Maharashtra
2	ICAR- Central Institute for Jute and Allied Fibres	Kolkata	West Bengal
3	ICAR-National Rice Research Institute	Cuttack	Odisha
4	ICAR -Central Tobacco Research Institute	Rajamundry	Andhra Pradesh
5	ICAR- Directorate of Groundnut Research	Junagarh	Gujarat
6	ICAR-Indian Institute of Oilseeds Research	Hydereabad	Telangana
7	ICAR- Directorate of Rapeseed-Mustard Research	Bharatpur	Rajasthan
8	ICAR-Indian Institute of Rice Research	Hyderabad	Talangana
9	ICAR-Indian Institute of Seed Science	Mau	Uttar Pradesh
10	ICAR-Indian Institute of Millets Research	Hyderabad	Talangana
11	ICAR-Indian Institute of Soybean Research	Indore	Madhya Pradesh
12	ICAR-Indian Agriculture Research Institute	Pusa	New Delhi
13	ICAR-Indian Agriculture Research Institute	Hazaribagh	Jharkhand
14	ICAR-Indian Grassland and Fodder Research Institute	Jhansi	Uttar Pradesh
15	ICAR-Indian Institute of Maize Research	Ludhiana	Punjab
16	ICAR-Indian Institute of Agricultural Biotechnology	Ranchi	Bihar
17	ICAR-Indian Institute of Pulses Research	Kanpur	Uttar Pradesh
18	ICAR-Indian Institute of Sugarcane Research	Lucknow	Uttar Pradesh
19	ICAR-Indian Institute of Wheat and Barley Research	Karnal	Haryana
20	ICAR-National Bureau of Agricultural Insect Resources	Bangaluru	Karnataka

Sl.No.	Name of Research Institute	Location	State
21	ICAR-National Bureau of Agriculturally Important Micro-organism	Mau	Uttar Pradesh
22	ICAR-National Bureau of Plant Genetic Resources	Pusa	New Delhi
23	ICAR-National Centre for Integrated Pest Management	Pusa	New Delhi
24	ICAR- National Institute of Biotic Stress Management	Raipur	Chhatishgarh
25	ICAR-National Centre on Plant Biotechnology	Pusa	New Delhi
26	ICAR-Sugarcane Breeding Institute	Coimbatore	Tamil Nadu
27	ICAR-Vivekanand Parvatiya Krishi Anusandhan Sansthan	Almora	Uttrakhand
B	**HORTICULTURAL SCIENCES**		
28	ICAR-Central Institute for Arid Horticulture	Bikaner	Rajasthan
29	ICAR-Central Institute for Subtropical Horticulture	Lucknow	Uttar Pradesh
30	ICAR-Central Institute for Temperate Horticulture	Srinagar	Jammu and Kashmir
31	ICAR- Central Island Agricultural Research Institute	Portblair	Andman and Nicobar
32	ICAR-Central Plantation Crops Research Institute	Kasargod	Kerala
33	ICAR-Central Potato Research Institute	Shimla	Himachal Pradesh
34	ICAR-Central Tuber Crops Research Institute	Trivendrum	Kerala
35	ICAR-Directorate of Cashew Research	Puttur	Karnataka
36	ICAR-Directorate of Floriculture Research	Pusa	New Delhi
37	ICAR-Directorate of Medicinal And Aromatic Plants Research	Anand	Gujarat
38	ICAR-Directorate of Mashroom Research	Solan	Himachal Pradesh
39	ICAR-Directorate of Onion and Garlic Research	Pune	Maharashtra
40	ICAR-Indian Institute of Horticulture Research	Bangalore	Karnataka

Sl.No.	Name of Research Institute	Location	State
41	ICAR-Indian Institute of Oil Palm Research	Pedavegi	Andhra Pradesh
42	ICAR-Indian Institute of Spices Research	Calicut	Kerala
43	ICAR-Indian Institute of Vegetable Research	Varanasi	Uttar Pradesh
44	ICAR-National Research Centre for Banana	Tiruchirapalli	Kerala
45	ICAR-Central Citrus Research Institute	Nagpur	Maharashtra
46	ICAR-National Research Centre for Grapes	Pune	Maharashtra
47	ICAR-National Research Centre for Litchi	Muzaffarpur	Bihar
48	ICAR-National Research Centre for Orchids	Gangtok	Sikkim
49	ICAR-National Research Centre on Pomegranate	Solapur	Maharashtra
C	**NATURAL RESOURCE MANAGEMENT**		
50	ICAR-National Research Centre for Seed Spices	Ajmer	Rajasthan
51	ICAR-Central Agro-forestry Research Institute	Jhansi	Uttar Pradesh
52	ICAR-Central Arid Zone Research Institute	Jodhpur	Rajasthan
53	ICAR-Central Research Institute for Dry land Agriculture	Hyderabad	Andhra Pradesh
54	ICAR-Indian Institute of Soil and Water Conservation	Dehradun	Uttrakhand
55	ICAR-Central Soil Salinity Research Institute	Karnal	Haryana
56	ICAR-Indian Institute of Water Management	Bhubaneswar	Odisha
57	ICAR-Directorate of Weed Research	Jabalpur	Madhya Pradesh
58	ICAR Research Complex for NEH Region	Shillong	Meghalaya
59	ICAR Research Complex for Eastern Region	Patna	Bihar
60	ICAR-Indian Institute of Soil Science	Bhopal	Madhya Pradesh

Sl.No.	Name of Research Institute	Location	State
61	ICAR-Indian Institute of Farming System Research	Meerut	Uttar Pradesh
62	ICAR-Indian Institute of Soil Science	Bhopal	Madhya Pradesh
63	ICAR-National Bureau of Soil Survey and Land Use Planning	Nagpur	Maharashtra
64	ICAR-National Institute of Abiotic Research management	Baramati	Maharashtra
65	ICAR- National Organic Farming Research	Gangtok	Sikkim
66	ICAR_ National Research Centre for Integrated Farming	Morihari	Bihar
D	**AGRICULTURAL ENGINEERING**		
67	ICAR-Central Institute for Research on Cotton Technology	Mumbai	Maharashtra
68	ICAR-Central Institute of Agricultural Engineering	Bhopal	Madhya Pradesh
69	ICAR-Central Institute Post Harvest Engineering and Technology	Ludhiana	Punjab
70	ICAR-Indian Institute of Natural Resigns and Gums	Ranchi	Jharkhand
71	ICAR-National Institute of Research on Jute and Allied Fibres Technology	Kolkata	West Bengal
E	**AGRICULTURAL EDUCATION**		
72	ICAR-Central Institute for Women in Agriculture	Bhubaneswar	Odisha
73	ICAR-Indian Agricultural Statistics Research Institute	Pusa Campus	New Delhi
74	ICAR-National Academy of Agricultural Research Management	Hyderabad	Telangana
75	ICAR-National Institute of Agricultural Economics and Policy Research	Pusa Campus	New Delhi
F	**AGRICULTURAL EXTENSION**		
76	ICAR-Directorate of Knowledge Management in Agriculture	Pusa Campus	New Delhi
G	**ANIMAL RESEARCH INSTITUTES**		
77	ICAR-Central Avian Research Institute	Bareilly	Uttar Pradesh
78	ICAR-Central Institute for Research on Buffaloes	Hisar	Haryana

Sl.No.	Name of Research Institute	Location	State
79	ICAR-Central Institute for Research on Cattle	Meerut	Uttar Pradesh
80	ICAR-Central Institute for Research on Goats	Mathura	Uttar Pradesh
81	ICAR-Central Sheep and Wool Research Institute	Avikanagar	Rajasthan
82	ICAR-Directorate of Poultry Research	Hyderabad	Telangana
83	ICAR-Indian Veterinary Research Institute	Bareilly	Uttar Pradesh
84	ICAR-National Bureau of Animal Genetic Resources	Karnal	Haryana
85	ICAR-National Dairy Research Institute	Karnal	Haryana
86	ICAR-National Institute of Animal Nutrition and Physiology	Karnal	Haryana
87	ICAR-National Institute of High Security Animal Diseases	Bhopal	Madhya Pradesh
88	ICAR-National Institute of Veterinary Epidemiology and Disease Informatics	Bangalore	Karnataka
89	ICAR-National Research Centre for Equines	Hisar	Haryana
90	ICAR-National Research Centre for Camel	Bikaner	Rajasthan
91	ICAR-National Research Centre on Meat	Hyderabad	Telangana
92	ICAR-National Research Centre on Mithun	Jharnapani	Nagaland
93	ICAR-National Research Centre on Pig	Guwahati	Assam
94	ICAR-National Research Centre on yak	Dirang	Arunachal Pradesh
95	ICAR-Project Directorate on Foot and Mouth Disease	Nainital	Uttrakhand
H	**FISHERIES INSTITUTES**		
96	ICAR-Central Island Fisheries Research Institute	Barrackpur	West Bengal
97	ICAR-Central Institute of Brackish water Aquaculture	Chennai	Tamil Nadu
98	ICAR-Central Institute Fisheries Education	Mumbai	Maharashtra
99	ICAR-Central Institute Fisheries Technology	Cochin	Kerala

Sl.No.	Name of Research Institute	Location	State
100	ICAR-Central Institute of Fresh Water Aquaculture	Bhubaneswar	Odisha
101	ICAR-Central Marine Fisheries Research Institute	Kochi	Kerala
102	ICAR-Directorate of Cold Water Fisheries Research	Bhimtal	Uttarakhand
103	ICAR-National Bureau of Fish Genetic Resources	Lucknow	Uttar Pradesh

Appendix 3: List of International Agricultural Research Institutes/Centres

Sl. No.	Name of Research Institute/Centre	Founded in	Location	Country
1.	International Rice Research Institute [IRRI]	1960	Manila	Philippines
2.	International Wheat and Maize Improvement Centre [CIMMYT]	1963	Mexico City	Mexico
3.	International Centre for Tropical Agriculture [CIAT]	1967	Cali	Colombia
4.	International Institute for tropical Agriculture [IITA]	1967	Ibadan	Nigeria
5.	International Potato Centre [CIP]	1971	Lima	Peru
6.	International Crop Research Institute for Semiarid Tropics [ICRISAT]	1972	Hyderabad	India
7.	International Centre for Agriculture Research in Drylans Areas [ICARDA]	1977	Aleppo	Syria
8.	West Africa Rice Development Association [WARDA]	1970	Bouake	Cote divoire
9.	International Plant Genetic Resources Institute [IPGRI]	1974	Rome	Italy
10.	Centre for International Forestry Research [CIFOR]	1992	Jakarta	Indonesia
11.	International Centre for Research in Agroforestry [ICRAF]	1991	Nairobi	Kenya
12.	International Centre for Living Aquatic Resources Management [ICLARM]	1977	Makati City	Philippines
13.	International Food Policy Research Institute [IFPRI]	1975	Washington DC	USA
14.	International Irrigation Management Institute [IIMI]	1984	Colombo	Sri Lanka
15.	International Livestock Research Institute [ILRI]	1995	Nairobi	Kenya
16.	International service for National Agricultural research [ISNAR]	1979	Hague	Netherlands

Sl. No.	Name of Research Institute/Centre	Founded in	Location	Country
17.	Asian Vegetable Research and Development Centre [AVRDC]	1971	Taiwan City	Taiwan [China]
18.	International Centre for Genetic Engineering and Biotechnology [ICGEB]	1987	Trieste	Italy

Glossary

Accession: A sample of a crop variety collected at a specific location and time; may be of any size.

Act: The Essential Commodities Act, 1955 (10 of 1955).

Adaptation: The fitness of a genotype to a particular environment.

Additional Declaration: A statement that is required by an importing country to be entered in a phytosanitary certificate and which provides specific additional information pertinent to the phytosanitary condition of a consignment.

Alien species: Related species.

Allele: Alternative form of a gene.

beings, animal life or parasitic on plant species.

Bio-control Agent: Any biological agent such as parasite, predator, parasitoid, microbial organism or self replicating entity that is used for control of pests.

Biodiversity: The totality of genes, species and ecosystem in a region or the world.

Biosafety: Safe application of modern biotechnology with regard to human health, animal health and environment.

Breeders' Exemption: Rights provided to use protected material as the basis to develop a new variety or for other research work; also called research exemptions or breeders' privilege.

Candidate Variety: A variety that has to be protected.

Center of diversity: Geographic region with high levels of genetic or species diversity.

Competent Authority: An authority notified by the Central Government from time to time by notification in the Official Gazette.

Conservation of biodiversity: The management of biodiversity to meet requirement of the present and future generations.

Consignment: A quantity of seeds, plants and plant products or any regulated article consigned from one party to other at any one time shipment and covered by a phytosanitary certificate, bill of entry of customs, shipping/airway bill or invoice.

Container: A box, bottle, casket, tin, barrel, case, receptacle, sack, bag, wrapper or other thing in which any article or thing is placed or packed.

Controller: A person appointed as Controller of Seeds by the Central Government and includes any person empowered by the Central Government to exercise all or any functions of the Controller under this Order.

Copyrights: The exclusive rights to reproduce, sell and distribute a work, prepare derivative works and display the work publicly.

Cosmopolitanism: Attitude of concern for world as whole, interest in universal principles; commonly contrasted with nationalism.

Cotton: Includes ginned cotton, cotton linters and dropping, tripping, fly and other waste products of cotton mill other than yarn waste, but does not include cotton seed or un-ginned cotton.

Cultivar: A cultivated variety (genetic strain) of a domesticated crop plant.

Cultivar: Cultivated variety of plants

Cultural Globalization: Trans country flow of culture.

Cultural Imperialism: Form of cultural hegemony enabling some states to impose worldview, values, and lifestyles on others.

Database Rights: A 'database right' is an intellectual property right given to a computer database.

Dealer: A person carrying on the business of selling, exporting or importing seeds, and includes an agent of a dealer.

Designated Inspection Authority: The authority notified by the Central Government from time to time through a notification to be published in the Official Gazette for the inspection of the plants grown in post entry quarantine facilities.

Distinctiveness: The variety must be distinguishable in at least one character from previously available varieties.

Domain Name: The names and words that companies designate for their registered Internet Web site addresses,

Domestic biodiversity: the genetic variation existing among the species, breeds, cultivars and individuals of animal, plant and microbial species that have been domesticated, often including their immediate wild relatives.

Ecological Globalization: Global protection of ecosystem from degradation and pollution.

Entry Point: Sea port, airport or land customs stations through which import is permitted under this order.

Essentially derived variety: A variety derived from the initial variety and having most of the characteristics of initial variety but clearly distinguishable from initial variety.

Ethnoscapes: Movements of people including tourists, immigrants, refugees and business travelers.

Ex situ **conservation:** A conservation of biodiversity alive outside of its original habitat or natural environment.

Example variety: A variety that is used for comparison of a particular character.

Exclusive Rights: The legal privileges that only a copyright owner has with respect to their copyrighted work.

Export: Taking out of India by land, sea or air.

Export: To take or cause to be taken out from any place in India to a place outside India.

Export-processing zones: Also free trade zones. Selected areas in industrializing countries marked by low taxes and tariffs, subsidized infrastructure, and exemption from some regulations, designed to attract foreign direct investment and stimulate growth

Extant variety: A variety available in India such as farmers' variety, variety of common knowledge or any other variety of public domain.

Extinction: The permanent loss of a species caused due to natural failure to adapt to environmental change.

Farmer: Any person who cultivates crops either by cultivating the land himself or through any other person but does not include any individual, company, trader or dealer who engages in the procurement and sale of seeds on a commercial scale.

Farmers' Exemption: Legal rights provided to farmers to save, use, exchange, share or sell his farm produce of a protected variety; also called farmers' rights or farmers' privilege.

Farmers' Rights: The legal rights provided to farmers to save, use, sow, replant, exchange, share or sell his farm produce including seed of a variety protected under Plant Variety Protection Act.

Farmers' Variety: A variety that has been developed by a farmer and used for commercial cultivation for several years.

Finance scapes: Global flow of money.

Form: A form appended to this Order.

Fruit: Any fleshy portion of the plant, that contains seeds, which is used for consumption, including seedless fruit both fresh and dry but does not include preserved or prickled or frozen fruits.

Gene bank: Various organizations where genetic diversity is maintained in living state.

Gene: The functional unit of heredity; the part of the DNA molecule that encodes a single enzyme or structural protein unit.

Genetic diversity: Variety of genes and genotypes found in a particular crop species.

Genetic Resources: The genetic material of actual or potential value.

Genotype: The set of genes possessed by an individual organism.

Geographical Indication: A geographical indication is a sign used on goods that have a specific geographical origin and often possess qualities or a reputation that are due to that place of origin.

Germplasm: Plants in whole or in parts and their propagules including seeds, vegetative parts, tissue cultures, cell cultures, genes and DNA based sequences that are held in a repository or collected from wild as the case may be and are utilized in genetic studies or plant breeding programs for crop improvement.

Global governance: Rules and institutions for managing and regulating actions or processes of global import.

Globalization: The process of trans border free flow of products, services, people, culture, technology, and finance.

Grain: Seeds intended for processing or consumption and not for sowing or propagation.

Horticulture nursery: Any place where horticulture plants are, in the regular course of business, produced or propagated and sold for transplantation;

Ideoscapes: Global spread of ideas and political ideologies.

IGO. Intergovernmental organization: Formed by and membership restricted to states. Examples: UN, NATO.

Immovable Property: Fixed types of properties which cannot move from one place to other. Such properties include land, buildings [house, flat, bungalow, farm house] and gardens.

Import Permit: An official document authorizing importation of a consignment in accordance with specified phytosanitary requirements.

Import: An act of bringing into any part or place of territory of Republic of India any kind of seed, plant or plant product and other regulated article from a place out side India either by sea, land, air or across any customs frontier.

***In situ* conditions**: The conditions where genetic resources exist within ecosystems and natural habitats.

***In situ* conservation:** A conservation in the original habitat or natural environment.

In situ: The conservation of biological diversity in the original location.

Indigenous Peoples: Groups held to be original residents of certain areas, especially illiterate groups under threat of displacement due to development.

Industrial Design: Any original shape, picture, or some combination applied to a useful article of manufacture. In other words, it refers to the design of the mass-produced products of our everyday environment, from sinks and furniture to computers.

Information Globalization: Trans border flow of knowledge, idea and information.

INGO: International nongovernmental organization

Inspection Authority: An authority notified by the Central Government from time to time or an officer of the Directorate of Plant Protection, Quarantine and Storage duly authorized by the Plant Protection Adviser (PPA) for the purpose of approval and certification of post-entry quarantine facilities and inspection of growing plants in such facilities in accordance with guidelines issued by PPA.

Inspector: An inspector of seeds appointed under Clause 12.

Intellectual Property Rights: The legal rights provided to an inventor to derive economic benefits from his invention/innovation.

Intellectual Property: The product/process/idea which is outcome of the brain of a person and can be used on commercial scale for benefit of human kind.

Issuing Authority: An authority as envisaged under Schedule-IV of this order or duly notified by the Central Government from time to time either generally or specifically for issuance of import permit.

Keystone species: A species whose loss from an ecosystem would cause a greater than average change in other species populations or ecosystem processes.

Kind: One or more related species or sub-species of crop plants each individually or collectively known by one common name such as cabbage, maize, paddy and wheat.

Landraces: A crop cultivar or animal breed that evolved with and has been genetically improved by traditional agriculturalists, but has not been influenced by modern breeding practices.

Mask Work: A mask work is a two or three-dimensional layout of an integrated circuit.

Mediascapes: Global distribution of media images that appear on our computer screen, in news papers, television and radios.

Misbranded: A seed with false label or label of another variety.

Moral Rights: Moral rights are a special extension to copyright that gives the originator of a copyrighted work rights over its use.

Movable Property: The property which can be easily shifted from one place to other. Such property includes animals, farm machines, furnitures, fixtures, gold, silver, diamond, and money.

Multiculturalism: Doctrine asserting value of different cultures coexisting within single society; globally, vision of cultural diversity deliberately fostered and protected.

NGO: Nongovernmental organization.

Notification: A notification published in the official Gazette and the expression "notifies" shall be construed accordingly.

Novelty: Newness of a variety. A variety which has not been grown for more than one year prior the application for registration.

Noxious Weeds: Any weed harmful or hazardous or unwholesome to human

Nursery: Any orchard, or any other place, facility, glass-house, screen house, utilized for raising plants.

Official Phytosanitary Certificate: A phytosanitary certificate in the format (reproduced as Schedule I) prescribed by the International Plant Protection Convention sponsored by the Food and Agricultural Organization of the United Nations Organization and issued by the authorized officer of the country of origin of consignments.

Orthodox seed: Seed that can be dried to moisture levels between 4 and 6 percent and kept at low temperatures.

Packing Material: Any kind of material of plant origin used for packing, which shall include hay, straw, wood savings, wood chips, saw dust, wood waste, wooden pallets, dunnage mats, wooden packages, coir pith, peat or sphagnum moss *etc.*

Patent: A document granting an inventor sole rights to an invention. It is an official document which grants sole rights to the inventor for manufacturing and marketing his product/process/invention to derive benefits.

Pest Risk Analysis: The process of evaluating biological or other scientific and economic evidence to determine whether a pest should be regulated and strength of any phytosanitary measures to be taken against it.

Pest: Any biotic agent capable of causing any injury or damage to plants and plant products and include any form or stage of insects, mites, snails, slugs, worms, nematodes, algae, fungi, protozoa, bacteria, actinomycetes, viruses, viroids and molecules and also include genetically engineered or modified organisms and weeds.

Phenotype: The external appearance of an individual.

Phytosanitary Certificate: A certificate issued in the model format prescribed under the International Plant Protection Convention of the Food and Agricultural Organization and issued by an authorized officer at the country of origin of consignment or re-export.

Plant Breeders' Rights: Plant breeders' rights [PBR], also known as plant variety rights (PVR), are intellectual property rights granted to the breeder of a new variety of plant.

Plant Product: An un-manufactured material of plant origin including grain and those manufactured products that, by their nature or that of their processing, may create risk for the introduction and spread of a pest.

Plant Protection Adviser: The plant Protection Adviser to the Government of India, Directorate of Plant Protection, Quarantine and Storage, N.H.IV, Faridabad.

Plant: Any plant or part thereof, whether living or dead, trees, shurbs, nursery stock, and includes all vegetatively propagated materials.

Point of Entry: Any sea port, airport, or land-border check-post or rail station, river port, foreign post office, courier terminal, container freight station or inland container depot notified as specified in Schedule-I or Schedule-II or Schedule-III as the case may be.

Post Entry Quarantine: Growing of imported plants in confinement for a specified period of time in a glass house, screen house, poly house or any other facility, or isolated field or an off-shore island that is established in accordance with guidelines/standards and are duly approved and certified by an inspection authority notified under this order.

Post-entry Quarantine: Growing of plants in isolation for any specified period in a glass-house, and facility, area of nursery, approved by the Plant Protection Adviser.

Producer: A person, group of persons, firm or organization who grow or organize the production of seeds.

Production Globalization. Trans country flow of goods or products.

Property: The wealth or valuable things earned by a person.

Public Domain: The knowledge that is freely available, commonly shared throughout the world without any access restrictions. It cannot be protected by intellectual property rights.

Quarantine Pest: A pest of potential economic importance to the area endangered thereby and not yet present there, or present but not widely distributed and being officially controlled.

Recalcitrant seed: Seed that does not survive drying and freezing.

Reference variety: All released and notified extant varieties that are under seed production chain

Registering Authority: A licensing authority appointed under clause 11.

Regulated Article: Any article the import of which is regulated by this order;

Schedule: Schedule annexed to this order.

Seed bank: A facility designed for the *ex situ* conservation of individual plan varieties through seed preservation and storage.

Seed processing: The process by which seeds and planting materials are dried, threshed, shelled, ginned or delinted (in cotton), cleaned, graded or treated.

Seeds: Seeds of agricultural and horticultural crops and forest plant species produced by sexual reproduction and shall include naked seeds (cones) produced by gymnosperms and seed sprouts meant for propagation or consumption. The seed as defined in the Seed Act, 1966 (54 of 1966).

Selection: The process that favors survival and further propagation of some plants having more desirable characters than others. Natural selection favors those traits which are essential for survival of a species. Artificial selection favors those characters which are of economic value to the human.

Soil: Earth, sand, clay, silt, loam, compost, manure, peat or sphagnum moss, litter, leaf waste or any organic media that support plant life and shall include ship ballast or any organic medium used for growing plants.

Species diversity: Variety of species found in a given region or area.

Species: A group of organisms capable of interbreeding freely with each other but not with members of other species.

Spurious seed: Any seed which is not genuine or true to type.

Stability: The same performance of a variety after repeated reproduction or propagation.

State Government in relation to a Union Territory: The Administrator hereto by whatever designation known.

Subspecies: A subdivision of a species.

Sui Generis Rights: Sui-generis refers to things of their own kind or things with unique characteristics.

Supplementary Protection Certificate: This term is used in European Countries. This is an extension of the term given to pharmaceutical or plant protection patent.

Sustainable Development: Policy of promoting growth consistent with protection of environment

Technoscapes: Movement of technologies around the world.

Timber: A form of dead wood, log and lumber cut from plants, with or without bark or sawn and sized, which is used for manufacturing veneer, plywood, particle or chip board and making building material, furniture, packages, pallets, sports goods and handicrafts.

Tissue Cultured Plant: Any part of a plant or plant tissue or plantlet grown under aseptic or sterile conditions in flasks or other suitable container on appropriate media and shall include ex-agar washed plant lets.

Trade Name: The name under which a company conducts business, or by which its business, product/service are identified. It may or may not be registered as a trademark.

Trade Secret: A formula or process or device used in business that is not published or divulged which gives an advantage over competitors.

Trademark: A trademark can be a word, name, symbol, device or mark which is used to identify and distinguish the goods or services of one company from goods or services of another.

Traditional Knowledge: The Traditional Knowledge [TK] refers to the knowledge, innovations and practices of indigenous people and local communities.

Transgenic variety: Seed or planting material synthesized or developed by modifying or altering the genetic composition by means of genetic engineering.

Uniformity: All plants in a variety should look alike.

Variety: A plant grouping within a single botanical *taxon* of the lowest known rank.

WTO: An international organization that regulates trade and tariffs between nations in order to ensure that trade flows smoothly, predictable and as freely as possible.

Key References

A. SEED LEGISLATIONS

1. Anonymous, 1966. Indian Seed Act 1966 [Act No. 54 of 1966]. Government of India Ministry of Agriculture (Department Of Agriculture), December, 29, 1966.

2. Anonymous, 1968. Seed Rules, 1968. Government of India Ministry of Agriculture (Department of Agriculture), 1968.

3. Anonymous, 1973. The Seeds (Amendment) Rules, 1973, Government of India Ministry of Agriculture (Department of Agriculture), No. 7(17)/69-Seeds Dev. New Delhi, Dated The 30.6.1973

4. Anonymous, 1983. The Seeds (Control) Order, 1983. Under seed Act, 1966 (Act No. 54 Of 1966). Government Of India, Ministry Of Agriculture, (Department of Agriculture and Cooperation), New Delhi, Dated The 30th Dec. 1983.

5. Anonymous 1988. New Seed Policy. Ministry of Agriculture, vide Letter No.11-71/88-SD-1 dated September 16,1988.

6. Anonymous, 1989. The Plants, Fruits and Seeds (Regulation of Import into India) Order, 1989 Ministry Of Agriculture (Department of Agriculture and Cooperation), New Delhi, the 27th October, 1989.

7. Anonymous, 2001. The Protection of Plant Varieties and Farmers' Rights Act, 2001, Ministry of Law, Justice and Company Affairs (Legislative Department), *New Delhi, the 30th October, 2001.*

8. Anonymous, 2002. National Seed Policy, 2002. Government of India, Ministry of Agriculture (Department of Agriculture), 19th June, 2002.

9. Anonymous, 2002. The Biological Diversity Act, 2002, Government of India, 2002.

10. Anonymous, 2004. Seed Bill 2004. Government of India, Ministry of Agriculture (Department of Agriculture), March, 2004.

11. UNIVERSAL, 2006. The protection of Plant Varieties and Farmers' Rights Act, 2001 and The protection of Plant Varieties and Farmers' Rights Rules, 2003. Bare Act with Short Notes. Universal Law Publishing Co. Pvt. Ltd., New Delhi.

12. Singh Phundan 2013. Essentials of Plant Breeding, 5th edn. Kalyani Publishers, New Delhi.

13. Singh Phundan, 2009. IPR and Plant Breeders' Rights [Subjective]. New Vishal Publications, New Delhi.

14. Singh Phundan, 2010. IPR and Plant Breeders' Rights [Objective]. New Vishal Publications, New Delhi.

15. Singh Phundan, 2010. IPR and Plant Breeders' Rights [At a Glance]. New Vishal Publications, New Delhi.

B. PLANT BREEDING

1. **ALLARD, R. W.** 1960. Principles of Plant Breeding. John Wiley and Sons Inc. New York.

2. **ALLARD, R. W.** 1999. Principles of Plant Breeding 2nd edn. John Wiley and Sons Inc. New York.

3. **BOROJEVIC, S.** 1990 Principles and Methods of Plant Breeding. Alsevier, Amsterdam.

4. **BRIGGS, F. N. AND KNOWLES, P. F.** 1967. Introduction to Plant Breeding. Reinhold, New York.

5. **CHOPRA, V. L. (ED.)** 2000. Plant Breeding: Theory and Practice 2nd edn. Oxford and IBH Publishing Company, Pvt. Ltd. New Delhi.

6. **FEHR, W. R.** !987. Principles of Cultivar Development Volume 1. Theory and Techniques. Macmillan Publishing Company, New York.

7. **FREY, K. J.** (ed.) 1966. Plant Breeding. Iowa State University, Press, Ames, Iowa, USA.

8. **FREY, K. J.** (ed.) 1981. Plant Breeding II. Iowa State University, Press, Ames, Iowa, USA.

9. **HARTEN, A. V.** 1998. Mutation Breeding: Theory and Practical Applications. Cambridge Uni. UK.

10. **KUCKKUCK, H., KOBADE, G. AND WENGEL G.** 1993. Fundamentals of Plant Breeding. Narosa Publishing House, New Delhi.

11. **MAYO, O.** 1984. The Theory of Plant Breeding. Clarendon, Oxford.

12. **POEHLMAN, J. M.** 1987. Breeding Field Crops 3rd edn. AVI Publishing Co. Inc. West Port, Connecticut, USA.

13. **POEHLMAN, J. M. AND BORTHAKUR, D. N.** 1969. Breeding Asian Field Crops. Oxford and IBH Publishing Company, New Delhi.

14. **PHUNDAN SINGH** 2010. Essentials of Plant Breeding 4th edn. Kalyani Publishers, New Delhi.

15. **PHUNDAN SINGH** 2009. Genetics 2nd edn. Kalyani Publishers, New Delhi.

16. **PHUNDAN SINGH** 2012. Cotton Breeding 3rd edn. Kalyani Publishers, New Delhi.

17. **PHUNDAN SINGH AND SANJEEV SINGH** 2010. Breeding Hybrid Cotton 3rd edn. Kalyani Publishers, New Delhi.

18. **SHARMA, J. R.** 1984. Principles and Practice of Plant Breeding. Tata McGraw-Hill Publishing Company Ltd. New Delhi.

19. **SIMMONDS, N. W.** 1979. Principles of Crop Improvement. Longman, London.

20. **SNEEP, J. AND HENDRIKSON, A. J. T.** (eds) 1979. Plant Breeding Perspectives. Pudoc, Wageningen, Netherlands.

21. **VOSE, P. B. AND BLIXT, S. G.** (eds) 1984. Crop Breeding A Contemporary Basis. Pergamon Press, London.

22. **WILLIAMS, W.** 1964. Genetic principles and Plant Breeding, Oxford, Blackwell.

C. GENETICS

1. **ACQUAAH GEORGE.** 2006. Principles of Plant Genetics and Breeding. Blackwell Publishers.

2. **CARROLL, M.** 1989. Organelles. Macmillan Press Ltd. Basing Stock, U.K.

3. **DOUCE, R.** 1985. Mitochondria in Higher Plants. Academic Press, Inc. Orlando, Florida, USA.

4. **FRIEFELDER, D.** 1987. Molecular Biology 2nd ed. Jones Bartlett Publishers, Inc. USA.

5. **GARDNER, E. J., SIMMONS, M. J. AND SNUSTED, D. P.** 1991. Principles of Genetics 8th ed. John Wiley and Sons Inc., New York, USA.

6. **HERSKOWITZ, I. H.** 1979. Elements of Genetics. Macmillan Publishing Co. Inc., New York, USA.

7. **KHUSH, G. S.** 1976. Genetics of Aneuploids 2nd ed. Academic Press Inc., New York, USA.

8. **SNUSTED, D. P. AND SIMMONS, M. J.** 2005. Principles of Genetics. John Wiley and Sons, Inc., New York, USA.

9. **PHUNDAN SINGH 2009.** Genetics 2nd edn. Kalyani Publishers, New Delhi.

10. **PHUNDAN SINGH 2011.** Elements of Genetics 3rd ed. Kalyani Publishers, New Delhi.

11. **STRICKBERGER, M. W.** 1985. Genetics. 3rd ed. Macmillan Publishing Co., New York, USA.

12. **SINGH, B. D.** 1995. Fundamentals of Genetics. 2nd ed. Kalyani Publishers, New Delhi

C. STATISTICS

1. **CHANDEL, S.R. S.** 1993. A Hand Book of Agricultural Statistics, 3rd ed. Achal Prakashan Mandir, Kanpur, India.

2. **DAS, M. N. AND GIRI, N. C.** 1986. Design and Analysis of Experiments, 2nd ed. Wiley Eastern Limited, New Delhi.

3. **ELHANCE, D. N. AND ELHANCE, V.** 1990. Fundamentals of Statistics. Kitab Mahal, New Delhi.

4. **GUPTA, S. K. AND KAPOOR, V. K.** 1993. Fundamentals of Mathematical Statistics. S. Chand and Sons, New Delhi.

5. **RAO, C. R.** 1952. Advanced Statistical Methods in Biometrical Research. Wiley and Sons, New Delhi.

6. **SINGH, S., SINGH, T. P., BANSAL, M. L. AND KUMAR, R.** 1991. Statistical Methods for Research Workers. Kalyani Publishers, New Delhi.

7. **PANSE, V. G. AND SUKHATME, P. V.** 1985. Statistical Methods for Agricultural Workers. ICAR Publication, New Delhi.

8. **FISHER, R. A.** 1970. Statistical Methods for Research Workers 14th ed. Oliver and Boyd, London.

D. BIOMETRICAL GENETICS

1. **FALCONER, D. S.** 1989. Introduction to Quantitative Genetics, 3rd ed. Longman, New York.

2. **HALLAUER, A. R. AND MIRANDA, J. B.** 1981. Quantitative Genetics in Maize Breeding. Iowa State Uni. Press Ames, USA.

3. **KEMPTHORNE, O.** 1957. An Introduction to Genetic Statistics. John Wiley and Sons Inc., New York.

4. **MATHER, K. 1949.** Biometrical Genetics. Methuem and Co. Ltd. London.

5. **MATHER, K. AND JINKS, J. L.** 1971. Biometrical Genetics. Chapman and Hall, London.

6. **SINGH, PHUNDAN AND NARAYANAN, S. S. 2013.** Biometrical Techniques in Plant Breeding, 4rd ed. Kalyani Publishers, New Delhi.

7. **SINGH, R. K. AND CHAUDHARY, B. D.** 1985. Biometrical Methods in Quantitative Genetic Analysis, 2nd ed. Kalyani Publishers, New Delhi.

E. SEED TECHNOLOGY

1. **AGRAWAL, P. K.** 1994. Principles of Seed Technology. ICAR, New Delhi.

2. **AGRAWAL, P. K. AND DADLANI, M.** (eds) 1987. Techniques in Seed Science and Technology. South Asian Publishers, New Delhi.

3. **AGRAWAL, R. L.** 1980. Seed Technology. Oxford and IBH Publishing Co. New Delhi.

4. **BOYCE, K.** 1981. Introduction to Seed Science and Technology. ICARDA, Aleppo, Syria.

5. **CHALAM, G. V., SINGH, A. AND DOUGALS, S. J. F.** 1967. Seed Testing Manual. ICAR and USAID, New Delhi.

6. **COPELAND, L. O. AND MCDONALD, M. D.** 1985. Principles of Seed Science and Technology. Burgess Publishing Company, Minneopolis, Minnesota, USA.

7. **DAHIYA, B. S. AND RAI, K. N.**(eds) 1997. Seed Technology. Kalyani Publishers, New Delhi.

8. **DOUGHLASD, J. E**. 1976. Seed Certification Manual. National Seeds Corporation and Rockefeller Foundation, New Delhi, India.

9. **JOSHI, A. K. AND SINGH, B. D.** 2004. Seed Science and Technology. Kalyani Publishers, New Delhi.

10. **KHARE, D. AND BHALE, M. S.** 2000. Seed Technology. Scientific Publishers, Jodhpur.

11. **NSC and RF,** 1972. Field Inspection Manual. NSC and Rockefeller Foundation. New Delhi.

12. **NSC,** 1972. A Hand Book for Seed Inspectors. National Seeds Corporation, New Delhi.

13. **SINGH PHUNDAN 2010**. Essentials of Plant Breeding. Kalyani Publishers, New Delhi.

14. **ROBERTS, E. H**. 1972. Viability of Seeds. Chapman and Hall, London.

F. PLANT BIOTECHNOLOGY

1. **BAJAJ, Y.P.S.** 1994. Somatic Hybridization in Crop Improvement. Springer Verlag.

2. **BHOJWANI, S. S.** (ed.) 1990. Plant Tissue Culture: Applications and Limitations. Elsevier, Amsterdam.

3. **BHOJWANI, S. S. AND BHATNAGAR, S. P**. 1990. Embryology of Angiosperms. Vikas Publishing Co., New Delhi.

4. **BHOJWANI, S. S. AND RAJDAN, M. K.** 1996. Plant Tissue Culture: Theory and Practice. Revised Edition. Elsevier, Amsterdam.

5. **CHAWLA, H. S.** 2002. Introduction to Plant Biotechnology. 2nd ed. Oxford and IBH Publishing Co.PVT. LTD., New Delhi.

6. **KHANNA, V. K.** 1999. Plant Tissue Culture. Kalyani Publishers, New Delhi.

7. **PHUNDAN SINGH** 2006. Introduction to Biotechnology. Kalyani Publishers, New Delhi.

8. **PHUNDAN SINGH 2012.** Plant Biotechnology. Kalyani Publishers, New Delhi.

9. **PRAKASH, J AND PIERIK, R.L.M.** (eds.). 1993. Plant Biotechnology: Commercial Prospects and Problems. Oxford and IBH Publishing Co. Pvt. Ltd., New Delhi.

10. **RAJDAN, M. K**. 1993. An Introduction to Plant Tissue Culture. Oxford and IBH Publishing Co. Pvt. Ltd., New Delhi.

11. **SENSEN, C. W.** (ed.) 2002. Essentials of Genomics and Bioinformatics. Wiley-VCH.

12. **SINGH, B. D.** 2006. Plant Biotechnology. Kalyani Publishers, New Delhi.

13. **VASIL, I. K. AND THORPE T. A.** 1994. Plant Cell and Tissue Culture. Kluwer Academic Publishers, Dordrecht, Netherlands.

14. **ZAITLIN, M., DAY, P. AND HOLLAENDER, A.** 1985. Biotechnology in Plant Science: Relevance to Agriculture in the Eighties. Academic Press, N.Y.

Books by Dr. Phundan Singh

A. Plant Breeding

1. Essentials of Plant Breeding
2. Fundamentals of Plant Breeding
3. Plant Breeding: Molecular and New Approaches
4. Molecular Plant Breeding
5. Commercial Plant Breeding
6. Commercial Plant Breeding: At A Glance
7. Commercial Plant Breeding: Objective
8. Objective Science of Plant Breeding
9. Genetic Principles and Plant Breeding
10. Genetics and Plant Breeding: Comparative Analysis
11. Principles of Plant Breeding[For PG Students]
12. Plant Breeding: At a Glance
13. Plant Breeding [For Under Graduate Students]
14. Principles of Plant Breeding [For UG Students]
15. Practical in Crop Breeding
16. Practical and Numerical in Plant Breeding
17. Numerical Problems in Plant Breeding and Genetics
18. Glossary of Plant Breeding and Genetics
19. Breeding Crop Plants for Stress Resistance
20. Introduction to Resistance Plant Breeding

53. Genetics and Man
54. Genetics Today

D. Quantitative Genetics

55. Biometrical Techniques in Plant Breeding
56. Application of Biometrical Techniques in Plant Breeding [Hindi Edition]
57. Objective Quantitative Genetics
58. Quantitative Genetics: At a Glance
59. Quantitative Genetics
60. Introduction to Agricultural Statistics
61. Agricultural Statistics: Objective
62. Agricultural Statistics: At A Glance

E. Plant Biotechnology

63. Introduction to Biotechnology
64. Plant Biotechnology
65. Principles of Plant Biotechnology
66. Plant Biotechnology: At a Glance
67. Objective Plant Biotechnology
68. Fundamentals of Molecular Biology and Plant Biotechnology
69. Objective Molecular Biology and Plant Biotechnology
70. Molecular Biology and Plant Biotechnology: At A Glance

F. Intellectual Property Rights

71. IPR and Plant Breeders' Rights [Subjective]
72. IPR and Plant Breeders' Rights [Objective]
73. IPR and Plant Breeders' Rights [At a Glance]
74. Introduction to Intellectual Property Rights
75. Intellectual Property Rights: At A Glance
76. Intellectual Property Rights: Objective

G. Miscellaneous Books

77. Elements of Baby Corn
78. Baby Corn Farming for Higher Income
79. A Guide for Competitive Examinations of Agriculture
80. Objective General Agriculture for Competitive Examinations
81. Autobiography of Phundan Singh